Biosemiotics

Volume 17

Series editors

Aims and Scope of the Series
Combining research approaches from biology, semiotics, philosophy and linguistics, the field of biosemiotics studies semiotic processes as they occur in and among living systems. This has important implications and applications for issues ranging from natural selection to animal behaviour and human psychology, leaving biosemiotics at the cutting edge of the research on the fundamentals of life.

The Springer book series *Biosemiotics* draws together contributions from leading scholars in international biosemiotics, producing an unparalleled series that will appeal to all those interested in the origins and evolution of life, including molecular and evolutionary biologists, ecologists, anthropologists, psychologists, philosophers and historians of science, linguists, semioticians and researchers in artificial life, information theory and communication technology.

More information about this series at http://www.springer.com/series/7710

Gianfranco Marrone • Dario Mangano
Editors

Semiotics of Animals in Culture

Zoosemiotics 2.0

Springer

Editors
Gianfranco Marrone
Culture & Society
University of Palermo
Palermo, Italy

Dario Mangano
Culture & Society
University of Palermo
Palermo, Italy

ISSN 1875-4651 ISSN 1875-466X (electronic)
Biosemiotics
ISBN 978-3-030-10297-5 ISBN 978-3-319-72992-3 (eBook)
https://doi.org/10.1007/978-3-319-72992-3

Printed on acid-free paper

This Springer imprint is published by the registered company Springer International Publishing AG part of Springer Nature.
The registered company address is: Gewerbestrasse 11, 6330 Cham, Switzerland

Contents

Introduction: Towards Zoosemiotics 2.0

Gianfranco Marrone

Abstract This volume has at least three directions of discourse. The first is a dialogue between semiotics and other disciplines dealing in animal studies: philosophy, the philosophy of language and mind, the human and social sciences, the cognitive sciences and the neurosciences. The second is a discussion within semiotics, in the passage from zoosemiotics 1.0 to zoosemiotics 2.0. Last but not least is the cultivation of an empirical field of research, or rather a critical exploration of social culture in search of interesting phenomena or, rather, phenomena pertinent to our in-depth theoretical and epistemological study.

1 A Sort of Parable

Not so long ago, over Sunday lunch, I was conversing with a middle-aged lady on the subject of recent summer holidays. 'Don't get me started', she told me, 'I am a grandmother to dogs'. She explains that her two sons, neither of whom have a steady job, both of whom live with their girlfriends but are too scared to have children. As such, both couples have each bought a pair of large dogs, which breed I don't know. And they love them like crazy. During the holidays and on Saturday evenings, these dogs are wisely left at the parents' house. And by parents, I refer to the lady with whom I am sharing the table, who loves her children and everything they do, to the point of sacrificing her own, hard-earned holidays in order to look after the undeniably intrusive, but nonetheless dear animals. By now, they are pretty much considered part of the family, as has been the case in the USA for a long time now.

Here we have a short tale of a personal experience that sheds light on a multitude of emerging social phenomena and unexpected philosophical consequences. A sort of parable. What does 'a grandmother to dogs' mean? At first it seems like a simple expression, if not irreverent towards both grandmothers and dogs. Much like the situation she describes, which could be interpreted as a minor localised madness.

G. Marrone (✉)
University of Palermo, Palermo, Italy
e-mail: gianfranco.marrone@unipa.it

G. Marrone, D. Mangano (eds.), *Semiotics of Animals in Culture*,
Biosemiotics 17, https://doi.org/10.1007/978-3-319-72992-3_1

However, if we think about it, an entire world is opened up, an authentic anthropological condition, a psychological earthquake. On the one hand, the dogs' grandmother stands out against a backdrop of well-documented economic macro-phenomena (the financial crisis, growing unemployment, the increasingly precarious nature of employment) and their social consequences (giving up on the idea of marriage and being forced to give up on having children). On the other hand, what is interesting is the way in which this somehow leads to them being replaced (we'll put it like that) by widespread *bricolage*. As we know, often somewhat creative responses are developed in response to global problems, 'ways of doing' that were previously unthinkable, born of ulterior processes that are not simply metaphorical. So, as with all metaphors, we see substitutions that are, in fact, deferments being employed, translations that degenerate into betrayals: the dogs take the place of children without actually being them; the emotional relationship with the animal is a partial (and undoubtedly painful) transfiguration of the one between parent and child. In turn, the grandmother is no longer a grandmother in the same way, despite existing as a parental figure, with a pedagogical function and the offer of regular support. Even the dogs are, let's admit it, a little less 'dog' than before, given they occupy a space that has traditionally belonged to children, leading to the progressive (de)generation of a strange, new family and source of further chain reactions. For example, the entire universe of these gestures, the daily grind, the effort of family life, missing your unborn children destabilises life and yet, in its own way, makes it easier: no more heroic nappy changes and feeds at 3 am, the paediatrician is traded in for a trusted vet, whilst powdered milk, highly coloured rooms and plastic toys are cast aside for tasty biscuit treats, ridiculously comfortable beds and chains that aren't too tight round the neck *heaven forbid*.

To quote Bruno Latour (an author we will come back to later), one 'collective' is reserved in another, with everything that inevitably leads to a translation in terms of losses and gains, of loyalty and disloyalty. A 'collective', an articulated collection of actors who we dare to define as social and that, nonetheless, is made up of both humans and non-humans, men and beasts, spaces and things, sentiments and ideologies and institutions and processes: in short, reciprocal relationships.

From common sense, which is naively ontological, obsessively naturalistic, we move to a vision of the world that is systemic, structural and semiotic. We do not have humans on one side and animals on the other, which then strike up relationships with one another, relationships almost always of hierarchical subordination in which the former is the master and the latter is dominated. On the contrary, there is a specific relationship between the forces at play that tends to produce precise social actors, outlining their respective fragile identities: we have a new kind of parent, unwillingly deprived of the possibility of children, who turns their repressed affections to the dog, and we have a new kind of dog that, happily charged with being the substitute for economically unviable children, finds itself experiencing incredibly strong emotions, sharing spaces that are both entirely new and extremely comfortable. And so the human is no longer the dog's master, but his quasi-parent; and dog is no longer a domesticated beast or a faithful friend but a kind of child, the avatar of a baby. As for the grandmother, who is not only not the dog's master but nor its mother, she is perhaps precisely a quasi-grandmother who has a consequent emotional relationship with the dog, and we could, for instance, imagine how she will spoil it outrageously, endangering the most rigorous parenting strategies.

We need to be careful. We are not talking about the typical anthropomorphism of traditional folklore, fairy tales or Disney cartoons: there is no animal world that reinterprets a human one; there is no animal stereotype that behaves according to the corresponding human one. Nor has this story anything to do with the classic theme of hybrid figures, half human and half animal. Rather, what we find is a complex articulation of social and emotional relationships that generate *a brand new form of kinship*, a kinship in which the animal too plays a role, according to the subject with which it enters into the relationship: it will be the beloved child for the adoptive parents (we will use exactly this term here, adoptive), a dear grandchild for the adoptive grandmother, a suspicious sibling for the other animals in the family (dogs, cats, tortoises, birds?) and, why not, a cousin or in-law.

Now, in some cases, this might make certain hardened cynics smile, or scandalise many humanists. For others, it is an extremely interesting example of a relatively original socio-anthropological condition (relatively because this has been the case in the USA and elsewhere for quite some time), one that should be considered, interpreted and discussed. One possible interpretation, a minority one but convincing nonetheless, could, for example, indicate the growing diffusion within our society of a *widespread animism* that co-exists, in no strange way, with its dialectical opposite – naturalism. Animism, according to the anthropologist Philippe Descola (who we will also discuss later on), is the ontological position that thinks of humans and non-humans as being in psychological continuity and physical discontinuity. This is the opposite of naturalism, of course, which sees humans and non-humans as physically similar but incredibly diverse cognitively. To introduce the animal into a relationship of kinship is, for the animist, not an unexpected consequence but an ethnic necessity: be it the Amazonian Achuar people (who view apes as acquired relatives) or the activist who defends the last few seals left in the oceans (the activist thinks in scientific terms but acts accordingly to rigorously animist principles).

The animism present in our culture – the same culture that leads to a growing prevalence of animalism, vegetarianism, veganism and so on – is a new phenomenon that is also highly predictable. Unprecedented if we look closely at it using the tools of those farming out opinions for the media or the behavioural psychologist; taken as a given if observed from afar with the eyes of the anthropologist who tries, out of professional duty, to analyse the different human ethnicities spanning the millions of years they have been dispersed over the planet. Presenting deep similarities and feeble superficial differences. In one way or another, the relationship with the animal today requires further reflection, which must by definition involve scholars from diverse disciplines, authors, activists, journalists, ethologists and veterinarians, alongside politicians and administrators, companies and lobbyists and experts in marketing and communications. Today, *animal studies* is a fast-growing epistemological reality. Thinkers of all kinds are involved. It is worth involving also semiologists, who have always worked (through the branch known as zoosemiotics) on the sign systems and forms of communication of animals, and which today leads to a kind of *zoosemiotics 2.0*, more strictly dedicated to the analysis of the social culture of our times.

2 This Volume

As a result, this volume has at least three directions of discourse.

The first is a dialogue between semiotics and other disciplines dealing in animal studies: philosophy, the philosophy of language and mind, the human and social sciences, the cognitive sciences and the neurosciences.

Secondly, a discussion within semiotics, in the passage from zoosemiotics 1.0 (which has traditionally been more interested in the sign systems and forms of communication of various animal species) to zoosemiotics 2.0 (together with other sectors that might, at first sight, seem opposed, such as biosemiotics, on the one hand, and the semiotics of culture, on the other).

Last but not least is the cultivation of an empirical field of research, or rather a critical exploration of social culture in search of interesting phenomena or, rather, phenomena pertinent to our in-depth theoretical and epistemological study.

3 Towards Multinaturalism

Here we need to take a step back and trace out the theoretical background against which this problem can be considered using a *semiotics of culture* that knows how to get anthropological studies to interact with research on the media. The starting point is the conviction that behind all of this and many other current, similar standpoints when it comes to animals hides a more general change of epistemological position, both ethical and aesthetic, regarding a rather broader theme: that of nature. It would seem, in fact, that for some time here the traditional notion of 'nature' – and the social practices that descend from the same – is in a state of crisis.

Nature today is an *absolute value*: in politics, in tourism, in religion, in urban thought, in design, in consumption, in nutrition, in the media and so on. It is enough to add the adjective 'natural' to any one thing – be it a city or a snack or a fur or a detergent – to immediately provoke smiles and acquiescence, certainties and credit cards. It has become a kind of brand or perhaps a meta-brand. This absolute value has even pervaded (for very different reasons) the fields of human sciences, which are now more than ever advocates or an empiricism that has lost all uncertainty, increasingly in search of the so-called biological basis of the so-called human nature – among which is semiotics, which often follows the most widespread epistemological trends like a dog that is forever running behind, without really discussing them in depth, starting with its own internal significance: to what extent does the paradigm of signification oppose, as it should, that of raw, naked fact? Is meaning not contrary to the fact? Doesn't the primacy of the relationship not exclude, out of principle, any domain of objectivity?[1]

Today, the word 'nature' functions as an 'umbrella term': the semantic evidence for which it would like to provide a vehicle fails to hide a total lack of definition. For example, the growing epistemological naturalism doesn't necessarily correspond to the widespread concern for the health of the planet, respect for habitats and the conservation of that image of pure nature that lies behind every ecological and environmental ideology. Similarly, it is not a given that the issue of sustainability, so beloved of economists and designers alike, sociologists and urban planners, belongs on the same theoretical and symbolic plain as that of organic farming, highly sought-after by local and multinational companies in search of an added value to offer their customers on supermarkets shelves or to rich consumers filled with aesthetic anxiety about their health. And yet, they all use the term 'nature' (with its correlatives, synonyms and antonyms), despite the fact it may not correspond to the same concept, the same field of ideas and meaning, as we have an accumulation of many different things enclosed in the same word. The renewed high priests of truth – often perplexed custodians of good and evil, the beautiful and the ugly – are today accompanied (more by chance rather than by choice) by environmentalist followers of Gaia and lovers of the organic, promoters of lost paradises and business strategies for wellbeing.

And yet, as we often forget, this generalised option when it comes to nature as an environment to protect and a social value to defend – with all of the consequent overlapping and ambiguity – is the fruit of an epochal trend inversion. For millennia, human societies, especially their Western versions, have emerged and developed by going *against* nature, against the limits and constrictions that it would have imposed upon the human race. Progress – both mythical and real – is something that is born in opposition to a supposedly 'ungenerous' nature (Giacomo Leopardi *docet*), so that the sciences and technologies are configured much like other operations of discovery, even forced and violent ones, 'secrets' within which the supposed nature, slyly, entrenches itself. And now that this same progress, as our collective imaginary tells us, has surpassed every limit, destroying the environment and the planet, now it needs to reimagine itself, its own opportunities and its own results, perhaps going against that of centuries past and precisely towards a degrowth that happily allows it to recover a healthy relationship with the natural environment. Environmentalism promotes this degrowth.

It is well known that behind this most complex and varied of scenarios lies the entire history of Western society and culture, which laid its own political, epistemological, economic and religious foundations, building, undoing and reconstructing an articulation that is as weak as it is necessary, which creates a reciprocally constituted rapport between nature and society, divinity and humanity, objectivity and subjectivity and the I and the other. In short, facts and values. No political idea has been implemented without an idea of society that takes with it complementary ideas of nature, god and man. Above all, when negotiating that theoretical oxymoron that

[1] See at least Greimas and Courtes (1982), Fabbri (2008), and Marrone (2014b, 2017).

is 'human nature', whose contradictions have acted as a flywheel for ideologies and mythologies, axiologies and strategies. Much like a blanket that is too short, human nature is pulled towards a universal and eternal biological supposition (leading to the homogenisation of sociocultural differences in the name of a supposedly natural basis shared by all and metaphysical by principle) and then towards cultural articulations (leading to more or less strong differences between groups and individuals, societies and historical epochs).

However, today there are those who, aware of the very real fragility of these similar negotiations, even in the light of the current ethnic unrest caused in Western societies caused by mass migration, talk of *multinaturalism*, in an attempt to engrave that unfading conviction, at the same time both religious and epistemological, about the existence of a single, unique nature that provides the shared background for successive anthropological suppositions.

On one side we have Bruno Latour (1993, 1999, 2004, 2005, 2009, 2015) and those studying the so-called sociology of translation (Akrich et al. 2006), who have always maintained a dialogue with semiotics. They remind us that nature is the fragile result of scientific *discourse*: nature as realities is built from the scientific fare, it changes alongside it and it multiplies as science does, between continual, iridescent controversies in which science sits down with politics, administration, lobbies, religions, nationalisms and so forth. Controversies, therefore, are not simply scientific but also political. On the other, we have anthropologists such as Eduardo Viveiros de Castro (2009), who coined the term *multinaturalism* to describe the pronominal system of the Amazonian Achuar people, who – in simplified terms – use the pronoun *you* to describe a great many beings and entities that we Westerners would consider to be natural. Using Benveniste's formal apparatus of enunciation and Deleuze's principle of difference, Viveiros reconstructs an intersubjectivity that is extended to non-human subjects (animals, plants, dreams, etc.), an intersubjectivity which includes an *inter-objectivity* constructed differently from culture to culture, whereas in these cultures there is also the affirmation of a *plurality of nature*. It is Philippe Descola who, in *Beyond Nature and Culture* (2013, see also Descola 2010, 2011, 2014), casts a critical eye over the dualism nature/culture – drawn on extensively by semiotics, proposing a four-way partition of ontologies: alongside our *naturalism*, *animism*, *totemism* and *analogism*, which are four different ways of objectifying reality, placing it as other than itself, on the basis of the relationship of continuity or discontinuity between human psychic interiority and non-human exteriority.

4 Return to the Cave

We'll begin with Latour. His work takes as its starting point, particularly in the volume *Politics of Nature* (2004), the myth of Plato's cave, a pretext for the reconstruction of a profound nexus between politics and science, society and nature. The words used in Plato's text are of little importance; that which is fundamental, however, according to Latour, is the fact that this tale can be used as an event from which to roll out the figures and actors of an initial scene, rebuilding a script that is still performed in masks today. With this myth as our starting point, it is possible to demonstrate how the ideal division between nature and society is not a new issue but one that actually goes far back in time and, as such, is significantly more radical and complex than it first appears. Latour believes Plato's famous tale from the seventh book of *Republic* as particularly significant. In it, some slaves are chained up inside a cave whilst the sun shines outside; all they can see are the shadows of the things projected onto the rock face in front of them, and only the philosopher manages to free himself from his chains and see the real things outside the cave. In a single enunciative gesture, it establishes both a theory of a science and one of politics, starting with an essential spatial fracture, which is not so much a rupture between a dysphoric interior and a euphoric exterior, as we often hear, but a strong topical discontinuity, an effective *narrative disjuncture* between places and agents.

Latour observes that there are two contradictory aspects of Plato's tale. On one hand is the clear caesura: the chained prisoners do not know how to escape, and even when they are able to, they prefer not to; they stay inside and squabble, inert and ignorant, subjugated and foolish, the epitome of every 'hell of the social'. The real world, the heavenly reality of things, lies beyond them; it exists in spite of them and makes fun of them. This delineates a primordial schism between society, constitutively ignorant and as such incapable of giving itself rules for sensible co-existence, and nature, located elsewhere, whose entities exist autonomously mute. On the other, Latour goes on, the original caesura is experienced by the philosopher twice and in both directions. A fortunate escapee, the philosopher manages to see the reality of things, achieving both knowledge and truth; then he returns to the cave to tell the ignorant detainees how things really are. And if, in Plato's text, he meets a sticky end, not believed and subsequently lynched by the others, in later times he becomes a scientist; he will manage to give his own word to the world and impose, as a scholar, his own power in it. As a vessel of truth, he will put himself forward as the best possible governor, or at least the ideal advisor to the king. He will explain how things really stand, both inside and outside the cave.

From here comes the tacit, eternal alliance between science and power and between the truth of things (that exist in spite of man) and the human government (which doesn't need those things in order to act and impose its will). It is according to this ideal primitive division between nature and society that scientists and politicians, a priori, divide up their reciprocal expertise, helping each other, a posteriori, every time someone tries to attack their twofold dominion. The more the supposedly evident opposition between nature and culture is loudly proclaimed, with great condescension of the means, the more such a division is actually neutralised, put out of use or negated. This allows for the emergence not only of the culturally constructed character of every supposed natural reality but, above all, of a game that makes such a construction possible. And as soon as one attempts to point to all of this, to reconstruct the semiotic meta-mechanism (to use Lotman's words 1990) that lies at the basis of every cultural opposition between nature and culture, then one is accused of relativism (cognitive) and defeatism (social); one is reproached for not wanting to accept the real truth of the world (or science) and, with it, the good governance of human beings (or politics). The diarchy of the powerful and the scientist does not forgive: instead they use that which Latour calls the *rhetoric of the fist on the table* and reaffirm, without solving the problems of continuity, with arguments such as 'well, stop making life complicated; that's how things are, the facts speak for themselves; if someone doesn't believe in the laws of nature then they can throw themselves out the window from the eighth floor: in any case, as far as they are concerned the law of gravity is purely a social construct; surely they aren't scared?' This is followed by a sort of *fundamentalism of reason*, which evidence suggests is very similar to religious intolerance. It is no coincidence that today the accusation of relativism comes both from the scientific elite and the central institutions of major monotheistic religions. And so, two republican chambers are established and kept under close surveillance: science and politics, nature and society, non-human and human, truth and opinion, heaven and hell and objectivity and subjectivity. An authentic division of power, a political position (or perhaps a metapolitical one) that establishes what falls under the dominion of politics and what falls under the dominion of science. But there is also a scientific (or metascientific) division, an epistemological one, therefore, that decides what falls within the dominion of knowledge and what does not and belongs, therefore, to power. This is the result of the myth of the cave, a sort of participative double opposition: on one hand, a political meta-mechanism that distinguishes (to a lesser degree) that which is political from that which is scientific, and on the other, an epistemological meta-mechanism that distinguishes (to a lesser degree) that which is scientific and that which is political.

5 Facts and Values

Almost as a response, albeit oblique, to accusations of relativism that had already been posited by a number of scientific philosophers (Sokal and Bricmont 1998), Latour (2004) demonstrates here how the fundamental intention of his work (as is generally the case with a sociological-ethnographic approach to science) is in no way nihilistic. On the contrary, his work tends to reconstruct the political-epistemological dispositifs on which concrete scientific practices are founded and the philosophical theories that support it. Furthermore, it aspires to the constitution of 'a new shared world' or, rather, the reorganisation of the system – at once, political and scientific – of the two 'chambers'. From this we locate the conditions of possibility, of an authentic politics of nature.

Politics is defined first and foremost as 'the sum of tasks that allow for a progressive composition of a shared world', where by 'shared world' we mean a form of totalisation, the constitution of a 'something' as an organic and concluded 'everything', which very clearly involves inclusions and exclusions, as we will see. According to Latour, in order to build a common world as a collective entity, it is not enough to bring together the two existing chambers of nature and society, established using the double platonic trick. It is also necessary to disarticulate them one by one from the inside, to deconstruct them into their essential components and to then re-articulate the single elements that have been discovered according to a different order, original in its formulation but that responds to effective practices according to which both science and politics have always operated. So, the new separation of powers" hoped for by Latour foresees an eccentric way of carving up jurisdiction and competencies, power and knowledge, whose constitution will be slow and progressive and without definitive certainties, without rationality to oppose irrationality, where everything will be negotiated between those that tend towards order and closure and those that tend towards disorder and openness. And above all, *that no longer opposes facts and values,* both facts and values are 'unpacked', demonstrating their intimately contradictory nature, before being then 'repacked' in an efficient way so that they form two new chambers. Let's start with facts. On the one hand, facts are *facts*, in the sense that they are 'constructed' and therefore the concrete outcomes of an operation of production, of a transitive action of transformation of the world that stockpiles heteroclitic material (think of Saussure's language) in view of a practical or theoretical objective. On the other hand, however, the use of the term 'fact' as a noun leads us to forget the work necessary to produce it, much like the question of difference of question between one fact and another: not all facts are constructed in the same way; they find themselves at the same starting point and are equipped with the same guarantees of existence and so on. Furthermore, viewing facts as events that are inert, obtuse and mute that sit there in their own supposed evidence means forgetting that they have been formulated by some theory which uses them as examples of itself, or as empirical manifestations of some abstract model, or as the models themselves or even as expressions of a prototype. It is only by starting from some *theoretical point of view*, often retroactively justified by them, that facts are constructed and then posited as a given – therefore receiving not only a form but a sense.

Similarly, there is an internal problematicity in values. Above all, Latour observes, in our Platonic-modern cultural configuration, values are defined negatively with respect to facts: they follow them. The must being presupposes the being that also wants, in principle, to contribute to the creation. Values try to reintroduce some form of humanity to the fact, without chipping away at the fact as it stands, contributing to a corroboration of its innate insistence upon existing (e.g. first comes cloning, and then we think about it). As a consequences, without any basic material, the moralists end up falling into a quest for basic universal ethics, so far from the minutiae of everyday life and, as such, from the detail of facts, their contribution to which would actually be incredibly useful. The universal ethic loses sight of practical morality, much like the notion of fact had forgotten the production of knowledge.

In one way, much like the other, both facts and values show two faces that are almost entirely opposite. In order to resolve the internal contradiction, Latour argues that it is necessary to keep these two physiognomies separate and rebuild them together, in twos, thus reformulating the political-epistemological system of the two chambers. Maintaining the separation between facts and values means surreptitiously allowing the deception of the continuous passages from one to the other (think of racism). Reshuffling the cards, however, has as its aim the avoidance of such fraud, returning to the fact all of the pride of its planning phase and, to the value, the efficacy of its material foundation. Let's have a look how this might be possible.

Facts, as we have seen, have two important properties:

1. The ability of an entity to alter the order of the discourse, to interfere with habits, to surprise, to provoke perplexity and to complicate matters. Facts are born when beings of uncertain status ask to be considered; they put forward their candidacy to existence, demanding to be considered.
2. Evidence, the objectivity that closes the discussion, the outcome of an acclaimed acceptance of beings before they have manifested themselves, the institution that forgets the past and the end of discussions.

Values, in turn, have two characteristics:

1. They cause other, unconsidered entities to emerge; they pose the demand of a consultation to see if and whether to admit them to the collective, evaluating the possibility of adding other voices to the call.
2. In the concept of value, there is an almost opposite thrust, that is, to determine the choice, to push for the decision on what should happen and to take the decisions, therefore instituting hierarchies between what is right and what is not, between what is more correct and what is less correct.

In other words, in the old chamber of facts lies a need for *perplexity* and one for *institution*; in the old chamber of values, there is a need for *consultation* and another for *hierarchisation*. In the two new chambers, which Latour respectively calls of *Take into Account* and of Put in *Order*, these four elements find their place in pairs. Perplexity and consultation are located in the chamber of Take into Account (answering the question, how many of us are there?), whereas hierarchisation and institution are ordered within the chamber of Put in Order (answering the question, can we live together?). From this we can draw up the following framework:

1. Power to Take into Account 'How many of us are there?'
1.a. First demand (already inscribed in the concept of fact): 'You will not simplify the number of propositions that must be considered in the discussion' = *perplexity*.
1.b. Second demand (already inscribed in the concept of value): 'You will ensure that the number of voices to participate in the articulation of the propositions will not be circumscribed' = *consultation*.
2. Power to Put in Order: 'Can we live together?'
2.a. Third demand (already inscribed in the concept of value): 'You will discuss the compatibility of the new propositions with those already instituted, in such a way as to maintain all of them in the same shared world that will confer upon them their rightful place' = *hierarchy*.
2.b. Fourth demand (already inscribed in the concept of fact): 'Once the propositions have been instituted, you will no longer discuss their legitimate presence at the heart of collective living' = *institution*.

6 Political Fantasies and Mad Cows

Let's try and follow in a little more detail the process that articulates these four positions, bearing in mind certain issues. First of all, the process is circular: it can be started at any point. Secondly, it should be remembered that the various stages of this political-epistemological process derive from a recomposition of the implicit elements present in facts and values, de-structuring them and neutralising their opposition. Such stages, considered from positions that are, for the most part, scientific or scientific-political (such as, e.g. ecology), can be extremely useful when reflecting on other circumstances, such as those of social politics (think of the consequences of mass migration) or, as is our case, of that liminal sphere between nature and culture and science and politics that is animality. Finally, it should be noted how, in neutralising the difference between facts and values, the model has its own dose of useful ambiguity as it often passes without problem – or scandal – from the descriptive to the prescriptive, from being to having to be, from formal principle to substantial concreteness.

Let's return to the new subdivision of the two chambers as considered by Latour. The 'high chamber', called Take into Account, poses a numeric problem: how many propositions and how many entities must we take into consideration in order to articulate the shared world in a coherent way? The theoretical enemies, in this case, are scientific reductionism and political exclusion – as aprioristic as they are difficult to kill off. In the 'high chamber', there is no desire for an arbitrary or a priori reduction in the number of entities put forward. There is a kind of thrust towards the internal realm of the subjects who would like to be taken into account and whose candidacy gives rise to perplexity. This is what Latour calls *need for reality* – in which *recalcitrance* rather than 'reality' is the evidence given here: the obstinate desire to be (precisely) taken into account. This is followed by the activation of a consultation in order to open political contests or scientific controversies around these emerging subjects, around the perplexity they cause and around the aspirations they harbour. Consultation must be broad and must not be rushed so that it can measure the importance of the candidates. From here comes the opening of a series of inquiries looking for trustworthy witnesses, certified opinions and credible spokespeople. It is that which Latour calls, structurally speaking, *need for jurisdiction.*

The 'lower chamber', known as Put in Order, poses the problem of the continual organisation and reorganisation of new and old actors, both human and non-human. From this emerges an issue of compatibility between the old and the new, in such a way that ensures the new does not supplant the old, or vice versa, so that the old leaves the new space to act. The resulting hierarchy, which comes from the consultation that has taken place in the first chamber, is explicit and motivated, but it is continually being negotiated. From this comes a *need for publicness* (here 'publicness' is opposed to the 'clandestine' but also to the opacity of the procedures). It can finally proclaim institutionalisation, the final decision on social order and the laws of nature, on the overall physiognomy of the shared world. From this comes a *need for closure.*

Let's look at an example that has a great deal to do with animality. In the early years of the new millennium, as you will remember, there was much discussion of *prions* with the explosion of the hotly debated political/scientific/ecological episode that was BSE or mad cow disease. Early on, the question was framed in terms of the Platonic cave: on one hand there was a nature to be discovered (what is the real cause of the disease?) and, on the other, a society to govern (how to safeguard public health?). From this came the question posed by the French President Jacques Chirac to the celebrated professor Darmont, expert in the field, on how much responsibility for the disease could be attributed to certain nonconventional proteins called prions. But Darmont's response broke with the Platonic structure. He simply said, 'I don't know', as he did not intend to take on a responsibility – by affirming anything regarding the supposed 'natural' role of that protein – that transcended his role as a scientist. Darmont made it clear with that response, which at first sight might seem reactionary, that the issue was not uniquely or simply scientific but also, and perhaps above all, political, economic, gastronomic, touristic, ecological, technological, entrepreneurial and so on. It was an issue that, as such, required an obligatory (and happy) journey through both chambers, *Take into Account* and Put in *Order*. First, if we think about it, prions not only left researchers perplexed but also the farmers, Eurocrats, consumers, animal feed producers, the cows and the politicians. Their candidature to existence generates a need for a broad consultation between biologists, who quickly place their expert knowledge at the disposal of the collective, often contradicting other witnesses or spokespeople such as veterinarians, farmers, consumers, butchers, officials and cows, calves and even sheep. The levels of alarm were high, and the only way to face them effectively was to no longer put forward the *ancien régime* that distinguished between facts and values, assuming, in a positive way, an order that came in course without being rushed. And it is, in any case, necessary to stabilise the controversy in order to return to some kind of everyday life in which new hierarchies are instituted between that which is important and that which is less so. What should win out? The tastes of consumers, the country's economy or the animal's wellbeing? Or perhaps the balance of nature or the status of the researching scientists (to one of whom, Stanley Prusiner, a Nobel prize has already been awarded for his discovery of prions)? The institution of a clear hierarchy, public and negotiated, meant deciding at the same time if and what needed to change, how the way in which we eat meat or feed cows needed to be modified, how to present countries such as France or the UK to the tourism market as countries that love the taste of great meat and so on. To express an opinion regarding the existence of prions was, therefore, at once both a political gesture and a scientific assumption. Darmont was implicitly suggesting that, at some point, someone had to decide whether, from the perspective of overarching social organisation, the risk of contracting the disease was relevant or not, pertinent or not. At some point someone would have to ask much more radical questions – both ethical and economic at the same time – regarding the number of possible deaths that would it 'make it worthwhile' – as Latour reminds us, 8000 people die each year on French motorways, a figure long institutionalised as an assumed risk given the positive force that those motorways are for French society. Deciding where the axe should fall between ethics and economics is a political act but also a scientific one: Latour believes it hypocritical to distinguish between the two roles. But the story doesn't end there. The process, as we have said, is circular, so that which is deliberately and publicly left outside the shared world, precisely in order for it to be able to claim at any moment its own candidature to existence and to want to try to enter into it. From this comes Latour's ulterior theoretical passage: that of the constitution of exteriority, of a *new exteriority*, which is no long aprioristically nature but rather the outcome of an explicit process of exteriorisation. Exteriority is that which it has been decided will remain outside but that has a continual possibility of being called upon, allowing for a dynamic reactivation of the entire process. One day, prions will have their say.

7 Four Ontologies

This is the point at which I would like to reflect on the work of anthropologist Philippe Descola (2013). He starts from the idea of elaborating *action strategies* as intermediary forms between depth and the superficial, profound universal and idiosyncratic structures of manifestation. Here we have four types, each disarticulated between themselves by differences and oppositions, starting with the different ways in which discontinuity and continuity, interiority and physicality, the psychic and physical planes and human and non-human are perceived (also animals and plants and even artefacts and various divinities). Descola also says that we strive to make judgement outcomes in identity between ourselves and the things around us.

First we have *animism*, typical of the Amazonian Achuar, but not exclusive to it. According to animism, animals and plants (as well as devices, artefacts, supernatural beings, divinities, spirits and various other presences) have a soul, they are people like us and they have forms of intentionality, action plans and an ability to reason, feelings exactly like our own. They are 'dressed' in another way, they have a different appearance to the human one, but they belong to society on an equal par with humans. They therefore have the same rights and laws. Animals and plants, for example, have relationships and therefore an alliance with humans. Apes, of course, are like our brothers and sisters in-law: they are our relatives but only up to a certain point. If necessary, we can hunt and eat them. There is, therefore, a level of appearance and a level of reality. Society is made up of humans and non-humans: this is the reason, as Latour would say, it needs to be reassembled once more.

Totemism is, instead, typical of Australian aborigines. According to the aborigines, there are spirits linked to different places, a kind of *genii loci*, which act as a model – mould for the creation of various totemic groups. These include living beings – regardless of whether they are human, animal or plant – with shared generic characteristics, the same morals or physical qualities, such as behaviour (slow, lively, resourceful, passive), temperament (choleric, calm, happy, melancholy), form (large, burly, slim, round) and consistency (soft, hard, flexible, rigid). It is therefore possible that a woman, a serpent, an insect and a shrub are all slim, calm and resourceful, meaning they belong to the same group. This is because they

Table 1 Descola's ontologies

	Interiority	Physicality
Animism	Continuity	Discontinuity
Naturalism	Discontinuity	Continuity
Totemism	Continuity	Continuity
Analogism	Discontinuity	Discontinuity

descend from the same totem, and so despite their difference of species, they are different manifestations and incarnations of the same totem/model.

As for *naturalism*, this is our dominant ontology, and it is a rather recent one that has come into being with the experimental sciences. According to naturalism, humans and non-humans have the same natural basis, the same physicality and the same materiality, but only humans have an interiority. This is a way of totalising the world as other to us, using the form of a unique and compact nature. This is Kant's famous 'cosmological idea', highly plausible as a normative idea for scientific knowledge, and therefore rather paradoxical in facts, the empirical and that which we experience. From this comes both the exploitation of a natural world that is not ours, for common use and consumption, and the parallel problem of having to defend and safeguard it, because it affects us (a little light ecology). Lastly we have *analogism*, typical of the Orient and China, as well as the Andes and our own Renaissance. Here, everything is different from everything else; we live in a mode of singularity, of radical difference. From this comes a sense of disorder, which we can overcome by proposing vague analogies, a jungle of similarities by which things, though diverse, resemble one another through determined characteristics, in a cross-referencing/deferment ad infinitum, and without specific criteria. One thing is like another because both are warm; this other thing is then like a third because both are female; another two are similar because they are dry; and so on. The schema looks like Table 1.

Descola clarifies how every ontology has specific references, not necessarily in the same territories. These are ideal types that we can see throughout our own history, ways of objectivising and of perceiving continuity and discontinuity between ourselves and the world. A biologist (naturalist given their consideration of living things) may well consult a horoscope (analogism) or speak to his/her cat (animism). And so, rather than maintaining a distinct separation between these four ontologies, it will be rather more useful for us to see if and how they are interwoven together in our modern day, giving rise to extremely diverse forms of thought and behaviour based on possible hierarchies and the relative dominance at times of one ontology and then another.

I have an exemplary case that is more than a simple example because, according to Kuhn (1977),[2] it could be a thought experiment (*gedankenexperiment*) for us. It regards *natural* or *organic* wine, which given its name makes clear the evident problems in the problematic connection between facts and values. Natural wine, to use

[2] See also Goodman (1978, 1988).

Latour's conceptualisation, is the classic *entity called into question* that demands to *enter the collective.* This poses the problem of it eventually being *taken into consideration*, or determined as an entity in itself, different and autonomous from conventional wine, on the side of oenologists, wine critics and consumers. *Perplexities* arise regarding its value, with all the imaginable reactions ('it smells funny', 'it's undrinkable', etc.) but also its positive vindications, not just ontological but, as we will see, also ethical. Constructing it as an entity requires a change in the ethics of its production and of it being *made* as a fact. It sets forth the problem of its *institution* (at a certain point, this becomes part of the wine system, with labels and brands, from this, e.g. comes the difference between 'natural' – a generic term that is not branded – and 'organic', a term that is branded, the fruit of a precise form of recognition with stamps, guidelines and so on). Finally, there is the question of *hierarchy* (is natural wine more or less tasty than conventional wine?), thus leading to the reorganisation of the entire wine industry.

The trend for organic wine is a not indifferent tile of the complex mosaic of gastronomy or, rather, of 'gastromania' (Marrone 2014a). At first sight, the expression 'natural wine' seems an oxymoron: there is nothing more artificial than wine, nothing further from the idea of nature as a separate reality that exists in spite of the human. Hemingway said that wine was one of the greatest signs of civilisation in the world. If this phrase has any meaning, it is that which becomes clear when it is *placed in a narrative relationship with that which it opposes*: the conventional, if not industrial, wine that globalisation has made uniform both in terms of taste and production. A standardised taste, derived from the general demand to objectivise the quality of wine, corresponds to a production that has done all it can to achieve that taste, modifying the product according to the model to be maintained. And so, wines have ended up the same, regardless of who owns the vines and land or the traditions and styles of the winemakers: put plainly, regardless of *terroir.* In response to this planetary uniformisation comes the idea of a wine defined as natural, organic or biodynamic, depending on the rigour shown both in the vineyards and the wine cellar, exploiting the myriad variables 'naturally' (or rather, habitually, traditionally) faced in wine production. On one hand, then, wine is understood as natural *because of negatives*: vines cultivated *without* the use of fertilisers, herbicides or pesticides and *no* yeasts added to the must, *no* filtration and so on. On the other, this trend towards elimination (or rather, this inversion of trend with regard to common viticulture and its use of chemicals) is not an empty gesture towards a nature that, working on its own, would obtain better results. In fact, it is accompanied by the *value of diversity*, biological and anthropological, and therefore to the multiplication of possibilities within wine production and, with these, the possible quality and plurality of tastes derived from it (*analogism*?).

By refusing the idea of an objectification of the taste of wine, we open ourselves to a host of possibilities both regarding the moment in which we taste the wine, which no longer searches for a single prototype, and only one of quality, and, most importantly when it comes to production, a taste that must no longer follow pre-established standards, favouring the potentialities of every single *terroir*, and therefore bio- and ethnodiversity. In this way nature becomes plural, in a sort of *totemism* that doesn't know how to be such: *terroir* is a kind of totemic model from which vines and vineyards, cellars and oenologists and farmers and businesses are instituted.

Natural wine, you will have noticed, is not a thing, a product or a finished entity but the long and complex process that has brought it into being. We understand why such strong passions, both positive and negative, are inspired when it comes to natural wine. There are those who rail against the destruction of the age-old art of wine-making: organic wine, they say, gives off a bad smell. There are also those who, on the contrary, make natural wine their mission, if not religion, which when poured into the glass shows existence in its complexity, in the name of values such as honesty, mutual respect, equilibrium and universal harmony. From the aesthetic comes the ethical: there is an entirely ethical dimension to nature that is constructed around an ethical-political consideration of 'other' beings which, as Latour would say, are not represented in the parliament and do not have adequate rights or duties. Natural wine poses the problem of the ethics of things, of the vine and of the wine cellar before that of human beings: here we have an *animist* dimension of *deep* ecology or *anima mundi*, almost *new age*, in which the problem is posed regarding the rights of non-humans, thus upturning the Kantian mechanism of man as an end and not a means.

8 An Unsteady Threshold

It was worth pausing over these theoretical and epistemological presuppositions of that which we would like to call 'zoosemiotics 2.0', from which the question of animality can be reinterpreted using a semiotic and anthropological framework. Semiotising the animal – telling of it, saying it, representing it, thinking it, making it an object of science and knowledge – means managing a politics of the unsteady threshold: that which separates the animal, more or less obstinately, from the human being, first guaranteeing it a supposedly natural autonomy, one that is apparently free from ethical-political implications, and then undergoing continual incursions, on different levels and with different intensities, of the animal-human. To define the animal is to define the human, of course. What is more interesting, however, is observing how, often surreptitiously, we give an animal a place in the kingdom of 'nature', inserting it, in parallel, among the articulations of the social and among the institutions of culture (relatives, affections and affiliations, strategies and conflicts, etc.). In other words, making it an object of knowledge means demanding of it some political vision and praxis. In doing this, the current environmental, ecological and animalist demands, as well as vegetarian and vegan ideologies and ethics, have done nothing more than show us the Emperor's new clothes. There has always been a politics of animals: now it has meticulous and very determined spokespeople.

Semiotics, the science of systems and signification processes, does not escape this chiasma, which, from its first research strategies, inserted the (old) question of language in the different animal species among its foundational objectives. This was zoosemiotics, very in vogue in the 1950s and 1960s but equally lively over the following decades, and by the force of things, it found itself almost automatically marrying a paradigm of research both scientific and *naturalist*, arranging to keep the languages of humans (linked to the dynamism of history and culture) and that of non-humans (understood as more or less sophisticated codes of communication of an instinctual character) distinct. Montaigne's old warning remained in place, telling us it is not that animals don't talk but us who do not understand them as they use signs and not words, physical signals bestowed, perhaps, with nonlinguistic syntaxes.

The recent research discussed in the fields of the anthropology of nature and sociology of the sciences – brought together in the hypothesis of a *multinaturalism* renamed *ontological turn*[3] – allows for zoosemiotics to be considered in a new way. Rather than focusing on a presumed natural language of non-human animals, betting on their reflexive conscience, it would be more helpful, for a semiotics as a study of forms of human and social meaning, to work on their *discourse*, on the effective interactions, both communicative and scientific, and practical and functional, between humans and non-humans. These two groups, willing or otherwise, are reciprocally constituted, constantly reviewing this unsteady threshold that, in separating them, keeps them together. We will discuss all of this over the following pages.

References

Akrich, M., et al. (2006). *Sociologie de la traduction. Textes fondateurs*. Paris: Ecole des Mines.

Descola, P. (2010). *Diversité des natures, diversités des cultures*. Paris: Bayard.

Descola, P. (2011). *L'écologie des autres*. Paris: Quae.

Descola, P. (2013). *Beyond nature and culture*. Chicago: Chicago University Press.

Descola, P. (2014). *La composition des mondes*. Paris: Flammarion.

Fabbri, P. (2008). *Le tournant sémiotique*. Paris: Hermes.

Goodman, N. (1978). *Ways of worldmaking*. Indianapolis: Hackett Publishing.

Goodman, N. (1988). *Reconceptions in philosophy and other arts and sciences*. Indianapolis: Hackett Publishing.

Greimas, A.-J., & Courtes, J. (1982). *Semiotics and language: An analytical dictionary*. Bloomington: Indiana University Press.

Kelly, J. (Ed.). (2014). The ontological French turn. *HAU: Journal of Ethnographic Theory, 4*(1), 259–269.

Kuhn, T. A. (1977). *The essential tension*. Chicago: Chicago University Press.

Latour, B. (1993). *We have never been modern*. Cambridge, MA: Harvard University Press.

Latour, B. (1999). *Pandora's hope. Essays on the reality of science studies*. Cambridge, MA: Harvard University Press.

Latour, B. (2004). *Politics of nature. How to bring the sciences into democracy*. Cambridge, MA: Cambridge University Press.

Latour, B. (2005). *Reassembling the social. An introduction to actor-network-theory*. Oxford: Oxford University Press.

[3] See Kelly (2014).

Latour, B. (2009). *On the modern cult of the Factish Gods*. Duke: Duke University Press.
Latour, B. (2015). *Facing Gaia*. Cambridge: Polity Press.
Lotman, Y. M. (1990). *Universe of the mind: A semiotic theory of culture*. New York: Tauris.
Marrone, G. (2014a). *Gastromania*. Milan: Bompiani.
Marrone, G. (2014b). *The invention of the text*. London: Mimesis International.
Marrone, G. (2017). Farewell to representation: Text and society. In C. Bankov & P. Cobley (Eds.), *Semiotics and its masters* (Vol. 1). Boston-Berlin: De Gruyter.
Sokal, A., & Bricmont, J. (1998). *Intellectual impostures*. London: Profile Books.
Viveiros de Castro, E. (2009). *Métaphysiques cannibales*. Paris: PUF.

Part I
Animal as Food, Food for Animals

When to Eat Meat? Toward a Diet of Caring

Nicola Perullo

Abstract With this text I would like to suggest the idea that vegetarianism, understood as a "principle choice" based on rational arguments and ethical norms, is supportive of and in line with the philosophy that it usually intends to contend, that is, the individualistic and anthropocentric subjectivism that is typical of the most striking tradition of Western metaphysics. Since I do not believe in the goodness and effectiveness of this model and tradition, I intend to point out that classic, rigid vegetarianism is an ontological mistake that would be best avoided, precisely in order to promote a different type of thinking that is more inclusive, open, dynamic, and enlightened. I propose to name this strategy *gustatory wisdom*, based on a diet of caring and attention rather than on principles and rigid norms.

1 Life Is Not Given: Toward a Functional Diet

With this text I would like to suggest the idea that vegetarianism, understood as a "principle choice" based on rational arguments and ethical norms, is supportive of and in line with the philosophy that it usually intends to contend, that is, the individualistic and anthropocentric subjectivism that is typical of the most striking tradition of Western metaphysics. Since I do not believe in the goodness and effectiveness of this model and tradition, I intend to point out that classic, rigid vegetarianism is an ontological mistake that would be best avoided, precisely in order to promote a different type of thinking that is more inclusive, open, dynamic, and enlightened. With this statement, I situate my writing far from the normative ethics framework, considering it instead a proposal of situational ethic.[1] At the same time, however, I would also like to suggest that favoring the regulation of the consumption of meat products – hence, a diet that may limit their consumption – is

[1] For a normative ethic position in the philosophy of food, also with respect to the vegetarian choice, see, for instance, Harald Lemke's writings: Harald Lemke, *Ethik des Essens*, De Gruyter, 2016, and *The Ethics of Taste* (www.haraldlemke.de/texte/Lemke_Gastrosophy.pdf). Other essays on ethics and meat consumption in *The Philosophy of Food*, 2012, ed. by David M. Kaplan.

N. Perullo (✉)
University of Gastronomic Sciences, Pollenzo, Italy
e-mail: n.perullo@unisg.it

G. Marrone, D. Mangano (eds.), *Semiotics of Animals in Culture*,
Biosemiotics 17, https://doi.org/10.1007/978-3-319-72992-3_2

possible and perhaps even desirable *at this time*, starting with a different reference model, driven by strategic motivations and not by principles or standards. Of course, there is a myriad of books and studies on the relations between animalism, meat consumption, subjectivity, anthropocentrism, ethics, and gender: as far as I know and to the extent of my present goals, I will employ them without discussing them explicitly; with some of them, actually, my point of view will be quite sympathetic, with others not so much.[2]

At the core of my argument lies a strong and binding idea, which may seem not directly related to the present topic but which, in fact, is essential for it: life *is* not, rather life *goes* and carries on. Life is movement. In this conception, the world is not a network of interconnected points, but a *meshwork* of lines entangled ab origine: the meshwork is the world we live in; hence the world is a fluid entanglement (Ingold 2013). Following Ingold's perspective, this ontology is a fluid ontology, made of tangles and constant constitution, evolution, and decay; hence it is an ontology that there *is* not. Further radical consequence: the differences between entities; between people, animals, things, and places; and between myself and the other are not categorical differences of essence. They are positional, functional, temporary crystallizations within fields of forces and redefinitions of forms, a plexus of continual growth, development, and decay of relations. Thus, identities are multiple, because it does not exist any hypostasis of "I" and "other," "subject" and "object." The things of the world – all of them – are not objects: they are actually knots, entanglements. Beyond the concepts of subject/object and of intersubjectivity, then, which sees the world as a totality of discreet and separate units, a fluid world without objects and fixed entities is disclosed here, where subjectivity is always *in-between*. Another consequence of this position: life (hence, humanity) is not a given, but an endless task. Life is not something given in advance, a datum, rather it is something that grows and develops incessantly. If we accept the idea that humanity is not a fact that is to be verified but a task to fulfill – according also to Raimondo Lullo and Ortega y Gasset – then nothing is less certain about the stability and fixity of the boundaries between "human" and "living being" in the global sense (Ingold 2015). If the identities – "human" and "animal" respectively – are not hypostatized, fixed, and demarcated once and for all, the "dilemma of the omnivore" (to quote Pollan's major work) thus takes on a different meaning. I would like to suggest setting the question of (not) eating animal flesh within a non-ontological, but actually functional and positional, perspective, addressing the problem of the diet on a different level. A fixed ontology of diet – *What should I eat?* once and for all – would leave the field to a radical relationalism: living beings are not shaped by stable essence but by the relationships they establish and the encounters they make. So, what must be then cared for is the position of a specific relationship in a particular field. A functional diet would therefore be a continual consideration, the always morphogenetically renewing care of the relationships between what is (has become) human and what is (has become) an edible living being.

[2] For instance, I think that I agree with Chrystina Van Dyke's position, discussed in "Manly Meat and Gendered Eating" (Van Dyke 2015).

2 Losing the Boundaries: Living Instead of Being

"What should we eat?" refers, sometimes even implicitly, to the question "*how* should we live?". Eating and living are of course interwoven: without food, there is no life. Life carries on because of the (quality and quantity of) food we intake. If we reverse the levels of modality and content of those questions, we find that "*How* should we eat?" prompts us to ask ourselves "*What* should we live?" that is, what we consider to be the content of life, even of the life that we eat. In order to eat, we must know what life is and, in order to live, we must know what to eat. All the questions that ask what the animal "is" presuppose the theme of the boundary, the definition and the difference with the human being. Since the very beginning of philosophy, many have proposed differential definitions of what "is" human and what "is" merely animal; nevertheless, numerous exceptions were excogitated, as several philosophers have realized that the problem of the animal lies precisely in the problem of definition and classification as such.

Let me briefly give an example of fluid identity and transformation through and with food. I was educated in philosophy firstly through Wittgenstein and Derrida, and then I proceeded with Vico: the problem of the animal has always crossed my mind thanks to them. Vico thought of the origin of the human language and mind through the hypothesis of "bestioni," coarse and uneducated human beings, who, like beasts, "feel without reason," have senses but no abstraction, have unstructured gestural writing, and the whole series of oppositions that are recast by the great modern and post-Cartesian debate on the soul of beasts (Perullo 2002). The animal is defined by contrast and by analogy with man: the more similar it is to the latter, the higher being it is considered to be. Do animals have a language? Do they just react or can they answer, too? Do they know how to laugh? Do they have a countenance or just a face? A hand or just a limb? Do they pass away or just simply die? Do they suffer like we do? All of these questions, the outcome of a great and profound philosophical tradition, are based on the idea of ontological recognition: identity and difference. This acknowledgment derives from an analogy strategy: primates are almost like us and mammals a little less, fish do not really resemble us, and algae are not similar to us at all. In this chain of oppositions and similarities, *what* and *how* to eat also come up. Do animals have the ability to taste? According to the nineteenth-century gastronome Brillat-Savarin, of course not; according to the zoosemiotic of contemporary taste, at least to some extent, some animals do. According to the general standard conception, animals just eat and do so mostly *in solitude*, while humans have meals *together*; it follows that conviviality – and a consequent kind of society – defines only humankind.[3]

"If a lion could speak, we would not understand him" is a very famous statement by Wittgenstein (Cyrulnik 1998). The Austrian philosopher perfectly pointed out that the problem of the animal and of animality concerns classification, definition, and thus translation – translatability. But can one really translate without postulating

[3] For a complete history of the topic, see De Fontenay (1998).

a demarcation between ontologically different entities? Likewise Derrida, like Montaigne before, on his *The Animal That Therefore I Am*, insisted on the look of his cat. Derrida related how he felt when his cat was staring at him in the bathroom as he stood naked: a feeling between shame and the thought that a cat is not able to intimidate. Is it a problem to be naked in front of an animal? This issue brought Derrida, through a convincing argumentation, to show well how the theme of "animal rights" was a full-fledged classical and anthropocentric motive, tending to ignore one disturbing aspect of this asymmetric relationship. Why are we not at ease in asymmetric relationships? In Derrida's philosophy, there is a thought about the animal that corresponds to his idea of poetry. Poetry is irreducible singularity, which is out of the definition and shatters every ontological closure (Derrida 2008).

I teach at an institution called The University of Gastronomic Sciences: its thematic and problematic focus is the issue of food, so the question of the animal has continued to come up for me, especially in terms of nutrition, as it is evident. I have many vegetarian students, but even for those who are not, reflecting on quality of animal and vegetable species is fundamental. But I think the issue of the animal has presented itself to me even more in a different way: food is the central issue at my University, but the disciplines that deal with it are varied and different. There is botany, chemistry, microbiology, technology, law, economics, zoology, history, anthropology, design, ecology, sociology, philosophy, etc. At Pollenzo, my identity as a philosopher has slowly faded because I have found myself coexisting with and sharing relationships in a purpose-based society – thinking of and practicing with food – made up of humans, certainly, but very different from each other because of approaches, methodologies, and sensitivity. Is this intellectual identity loss a bad thing? For some and for conventional academics, it certainly is; for me it is a good thing, and it is deeply relevant to the animal question, which is the question of definition, classification, and identity itself. If our identity is less rigid, then that of the nonhuman animal is so, as well.

What is the human/What is the animal: can we therefore avoid the ontological question? Can we have a positive relationship with the animal without falling into traps related to such questions like "does it have feelings and passions, and if so, to what extent?" or "since it suffers, does it have rights?" and others mentioned above? I believe it is possible and even highly desirable. How can we make it happen? By *living with* it. Why should this be desirable? The reason is the following: when we recognize two diverse, ontologically different entities, to then seek and find common traits and possibly establish a logic of respect, are we not perhaps laying the foundations for a vital paradigm that is exclusivist and hierarchical? Given that this logic is not inclusive, it excludes plants, certain forms of life, and certain animals. As we know, in the past we have often wondered whether "others" – the different, the distant – were human as *we are*. What do we establish as the base of this community, of this communion? What are the limits of the living? And so, what *is* life? The alternative I propose, following that minor paradigm of thought that suspects of all fixed ontology and of all systematic philosophy, is to conceive life as it actually happens, or as a production and continuous generation of forms that develop in the movement and the encounter, in the relationship. Then things develop, evolve, and change. If humanity does not *exist* but *occurs* and if the same applies to nonhuman living beings, how can we know what an animal or a human being is *regardless*? If life *is* not but it is *made*, if humanity is not a fact to be verified but a task to carry out, then the analogical principle, based on entities presumably defined as different and isolable, collapses. The question "how can we be assured that a cat *has* a consciousness, an intelligence, a soul?" has disappeared; at its place, we find an actual, real relationship with the cat – we make some things (or not) with her/him – intertwining each other's "identities" in the midst of the ongoing experience. Or, to answer the question above in a classical way, we can say: "Of course the cat possesses of this, (just) *because* he/she is looking at me."

I am aware that this can give rise to a paradoxical effect, which may seem deeply counterintuitive: how can we eliminate the undisputable difference between a cat or a cow and I? It is not a matter of eliminating differences, but of multiplying them infinitely. In fact, if we observe better what actually happens when a human and an animal meet, come into contact, and eventually live together, this apparent paradox will begin to disappear. Look at how this relationship goes without having an already fixed theoretical model! We can practice this "anthropology beyond humanity" (Ingold 2013) or a kind of "philosophy beyond knowledge" – knowledge intended as analysis and objectivation – by thinking of Donna Haraway's criticisms of gender/sex distinction: against the hypostasis of male/female binary ontology and the very idea that sex is dictated by nature and genre by culture. By overcoming the dichotomy of nature/culture, placing everything on a context of phenomenological immanence, we find *only* identities that are incessantly produced, in the same practices of acting on a minute-by-minute basis, through gestures, languages, and movements. Thus, we can conceive the idea of a fluid, positional, functional identity. "To Human is a verb," Ingold says, evoking the great vision of Lullo, Ortega y Gasset, but also of Bergson: humanity is a verb, it is a continuous process of *humanization*. The human being is therefore *animal hominificans*. It is about moving from an ontological view to an ontogenetic view. Everything is a flow (Ingold 2015).

If all this is clear, then we can also gain a new conception about sociality and conviviality: human/nonhuman relationships are not relationships between *different* societies. There is no "human society" and "animal society," not even in the plural (human societies and animal societies) since the society is one, a continuous process of hybridization and relationships, of knots and interweaving, between species and various entities. Society is not intraspecific but interspecific. Again, let us think about the experiences of everyday life: humans have continuous relationships with nonhumans, such as plants, animals, minerals, and things – why should the concept of society be limited to its intraspecific connotation? Only because of pure abstraction, we do confine the concept of society only to the human being community – on the basis of ex post classifications. Phenomenologically, we do have different relationships with stones, leaves, dogs, spiders, plastic bottles, umbrellas, cars, sea anemones, shoe soles, and awnings. Society *is* the complex of the whole world, indeed, a *cosmos*: a single society, which continually produces temporal and functional differentiations in its course. The model proposed here is a radical and oxymoronic "absolute relationalism": the bodies are not *in* relationship, but rather "are" *their* relationships (and therefore *are* not, as they constantly become). This is the transition from the conception of the relationship as interaction (action between two separate or separable entities of principle) to the one of the relationship as *correspondence*: entities are what they are because they gradually correspond to each other, they act and suffer, forming and differentiating slowly in the junctions and interstices as a whole (Ingold 2016).

3 Arguments Against Ontological Vegetarianism

Let us finally use the above argument for the vegetarian question. Following the same line of reasoning, we begin to observe that there is no "carnivorous food" per se just as there is no "vegetable food" per se. What a weird statement I am saying! What I am asking, however, is to look at things from a different perspective. This is so in many respects. On the one hand, firstly, meat food is already a hybrid in itself, since the animal (and what it *is* as such) eats vegetable; secondly, vegetable food is a hybrid because even plants are what they are (have *become*) as a result of crystallization of relationships with the rest – and so are what they "are" for the animals that are in a relationship with them, eating them but, more generally, living together with them. Food is thus never ontologically *pure*, animal or vegetable. In other words: on the one hand, we eat meat even when we believe we do not eat it. Derrida clearly understood this when he talked about "universal cannibalism," and also the Italian writer Italo Calvino has shown well in the novel "Sotto il sole giaguaro" (*Under the Jaguar Sun*) the figure of an inevitable cannibalism: we always eat meat because, metaphorically, we eat each other (but what is the ontological, definitive difference here between metaphorical and physical?). On the other hand, then, we eat vegetables even when we believe we do not eat them. An ontologist-someone for whom formal logic is the ultimate truth-would thus ask: is everything equal and impossible to differentiate, then? The answer is quite the opposite. Everything is incessantly and infinitely different and distinct. Distinctions, however, are not ontologically prefixed, and ethical choices cannot follow distinctions prepared and cooked in advance. Ethical choices rise from the "doing undergoing" flow (Ingold 2015), from the real situation.

The question of what to eat, and specifically whether (or not) to eat meat, can thus be circumvented or reformulated as a question of *how* to eat *life*, what is *living*: it becomes a question of circumstances, of situation, of *when*. When you eat another form of life – and since it is necessary to eat ("il faut bien manger") as Derrida, again, said – it is of essence to decide *what* life we will eat, from time to time.[4] I propose to call this strategic diet a diet of caring, as opposed to a diet of choice: if this latter one is modeled on principles and "a priori" norms (discussing in advance if it is "right/wrong" to eat animals, and then acting accordingly), the former one is carried and positioned again in a process that is never predetermined. The diet of

[4] See Perullo (2011) for a discussion on Derrida on eating.

caring is a diet-ethic wisdom without principles, which evolves and takes shape on a case-by-case basis with flexibility and by facing the motives that are encountered along the path of experience.

Vegetarianism, as a prescriptive and fixed dietetic model in all its variants – like ovo-lacto vegetarian, fish-vegetarian, raw, frugivore, etc. – as, of course, its apparent opposite, carnivorism, represents ontologizing visions that start from the idea of sharp and a priori distinction and separation between human and animal.[5] Within this paradigm, we can further identify three orders of arguments in favor of a vegetarian choice: whoever practices it can adhere just to one of these or two or all three (of course there are infinite variations and contaminations). We can briefly review them, pointing out their limitations from the perspective of the *diet of caring* we outlined above.

Animalist Argument. It is very old, has been relevant throughout the history of mankind, and is present in many ancient and modern philosophers – it is enough to recall Plutarch, Pierre Bayle's *Rorarius*, Jeremy Bentham, Tom Regan, and Peter Singer (for a review, see Fontenay 1998). At the very bottom of this argument is the idea of the injustice of killing the animal because the animal suffers, feels, and it is to some extent similar to human beings.

This argument, after Descartes's animal-machine notion, naturally had a new and powerful reappearance. If you adhere to the animalist argument, it may follow that since humans, as omnivorous beings, can choose what to eat, they can refrain from eating animal meat for the above reasons. I have already pointed out earlier the reason why this argument is weak. It presupposes in fact a hierarchical division between living entities, from which the classical objection of anti-vegetarians derives: a cow no, but algae yes? (For anti-vegetarians, this objection leads to the conclusion that you might as well eat everything all the time, but what I propose here goes in another direction.) What is edible is alive, so the logic of choice that would be rationally founded a priori is actually exclusive: respect for some forms of life would exclude others – and why? The explanation at this point often slides into the religious or spiritual level, thus bringing into play different cultural elements according to different times and places. But also from this point of view, are we sure that respect excludes sacrifice? In many communities, the killing of the animal does not rule out a profound and respectful relationship with it, on the contrary. Often, those who live with the animal will end up eating it. In rural communities, those who know the animal directly are never, or are only very rarely, vegetarian, unless there is a precept of faith (as in Hinduism) involved. In general, as I already pointed out, those who do not *thematize* the animal – *objectifying*, making it become an object of study – but who actually *live* and engage in daily relationships with it, do not follow an abstractly vegetarian diet; the choice of being vegetarian is rather most popular among intellectuals, philosophers, and artists who do not deal with animals, as well as for urban, industrial, or postindustrial citizens. The curious consequence in objectifying is to loosen (at least in appearance) the relational constraints

[5] A story of the Italian vegetarian movement, which is exemplary for these distinctions, is found in Capatti (2016).

with what is analyzed and studied: it is not the need for objectivity that has made vivisection possible; on the contrary it is the need to vividly vivisect and divide that feeds the objectifying and distancing predisposition. The logic of respect and the logic of sacrifice are not opposed: life as a relationship does not lead to the consequence of abstract and absolute irenicism, since agreements and disagreements, negotiations and conflicts, communing and variations are necessary one to the other. Respecting life does not mean, of course, not to consume it ever – that does not happen.

Health Argument. Also very ancient, it is often intertwined with the animalist argument, but it can also be considered on its own or in any case with prevalence, and this is precisely what happens today in rich Western communities. Since the Second World War, slowly after the invention of the Mediterranean diet, the consumption of meat – especially red meat – was considered the cause of some healty problems, mainly cardiovascular diseases and tumors. But even this topic, taken *sic et simpliciter*, encounters many limitations, as it always happens when health, illness, and medicine are discussed. In scientific communities, in the face of some totally vegetarian options, there is a wider tendency to contextualize and limit the subject. Two main reasons: (A) it is noted that the problem of a risk to human health is related to the *amount* of meat consumption, so this would not entail that meat is always bad, depending rather on quantity; (B) it also depends, and especially for some, on the *quality* of that food, in the sense that only meat raised in a specific way (as we shall see below with the ecological argument) would be detrimental to health. As it is evident, medicine linked to nutrition is more controversial and sometimes more contradictory than it is possible to imagine, and it is no longer just the classic case of the "French paradox" to indicate this. It is very difficult to resort to a rigid and definitive choice based solely on such topics. Finally, there is another seemingly provocative question that those who are free from established and trivial mechanisms of thought cannot help but ask themselves: What is meant by "health"? What is health? What does it mean to live well? The question is not abstract: the popular saying "It is better to live one day as a lion than one hundred years as a sheep" (a saying that, as we see, calls animals into question) goes right to the point. In general, it is not easy to determine what is "good for us" beforehand – of course, because good is a difficult concept to define a priori in absolute terms with respect to specific contents (which, as such, are variable: psychological, social, and biographical conditions can turn it around). Just as it is the case with life and human beings, the good *is not* but it *is made*. As Zen practice dictates, if any path can lead to illumination, any diet *in principle* (so to say, there are no principles at all) can lead to a healthy life.

Ecologist Argument. This topic is a rather new one and, among the three, is the one with the most recent history. Eating meat would be wrong, primarily or exclusively, on an ethical-political basis because intensive, industrial farming pollutes, is harmful to the environment, and produces many negative externalities. Raising animals costs much more energy than planting crops, especially when it is done mechanically and industrially. Martin Heidegger, in a famous and very controversial passage, expressed his disdain for this kind of production. Those who profess only this third argument are, however, "pseudo-vegetarians" or functional vegetarians; dietary choice is seen as an ethical, overall ecological-economic option, to be identified with an overall political strategy: "Vote with your fork" is a motto of contemporary gastronomic associations that goes directly to such commitment. This argument in favor of vegetarianism is certainly the least rigid and flexible and, in this sense, it is closest to a diet of caring: those who defend it have fewer objections to the principle in the face of a non-intensive small-scale breeding that is of low-impact and is respectful of natural cycles. It should be noted, however, that this sustainable meat production would be in any case more expensive than a vegetable farm; however the cycle remains within "acceptable" dimensions (but still, on what scale?), so meat consumption is allowed. However, even on ecological and environmental issues, there is a huge debate under way involving all fields of knowledge, with positions that are far from being unique and definitive. The "ecological crisis" is now a fact accepted by many, but its framework and the solutions that are proposed to overcome it are the most varied. The impact that the food issue may have on the environmental problem cannot be taken for granted and is subject to various interpretations. In the de-ontologizing perspective I propose here, in general, one should not separate human beings from the environment. We are the environment that we make for we are part of it, together with all other beings; the environment is not a reserve we occupy but a place we make. If there are no human societies and nonhuman societies, since society is one and the same, we can thus say that also society and the environment are consistent with each other.

4 Gustatory Wisdom as Diet of Caring

The reasons for opting for a vegetarian diet as a priori choice are based on a certain idea of value: value as in well-being and a healthy state, value as an ethical way of living, value as ontological and metaphysical commitment, value as taste preferences. Value is usually measured with a *quantum* of purity – moral, health, taste, political – that would indicate what is right and what is wrong. The being abstaining from meat eating would be a more fair, elevated being, even lighter, thus a more spiritual being. It is also interesting to note how this idea of purity – an abstinence from meat eating that is often accompanied by a similar ostracism toward alcohol and smoking, referred to as "vices," superfluities unnecessary to a healthy and beautiful life, and that are indeed harmful – has often had very particular political implications: if vegetarianism is sometimes linked to pacifism (e.g., we can think of the Gandhian movement), the cult of health and the body of the twentieth century has created unexpected alliances between abstinence from meat and movements that are linked to the cult of physical strength, such as fascism.

What I want to clarify now is what binds together the three arguments presented above: as the result of a logic of choice, they derive from the idea that human beings are mainly characterized by the ability to exercise a rational will. I can act rationally to impose the values that I have always rationally identified as being righteous. In this case, I can educate a fair taste because I *understand* what I have to do. This theory places life, the path of experience, and the flow (*the undergoing*) under the control of acting, doing (*agency/doing*) (Ingold 2015). The logic of caring acknowledges, on the contrary, that humankind is not exclusively, and not, above all, a rational entity, in which voluntary and intentional behavior prevails over circumstances, flow, and experience. It is enough to notice the following: the abstract intention of representing us humans as rational agents matches with the intention of treating the beasts, all of them, always a priori, as if they were human.

When animalist positions study and *analyze* the animal *per se*, in order to give it rights, they fall into an absolute anthropocentrism for which the other is like me, for I assimilate it regardless its specific relationships with me or someone else. Shifting from a logic of choice to a logic of caring, then, is like going from intention to attention. Intention puts the acting, seeking, judging, and grabbing subject at the center of action, who moves according to previously set goals. Attention, on the other hand, puts within the center of interest the action as such, the whole of what occurs in the midst of the experience: an "agencing" (Ingold 2016) without agent. In an attentive approach, the keywords are flexibility, feeling and wisdom, more than rules, understanding, and intelligence. The logic of the dietetics of caring is a situational logic, based on the "it depends." This opens up space for responsibility. To choose what to eat follows from flexible and always variable considerations of all correspondences and all the links that lead to a present experience and situation: it is a choice framed into caring, so to say without any fixed principles or rules, not vice versa. Responsibility precedes individual free will based on arguments promoted by intellectual inferences only.

Finally, I will give an example of the logic of care that specifically affects the diet. Some time ago, I met a young man on the train. I was reading a book about food and he was interested in it, so we started talking. He told me he was vegan, so I began to expose some of the remarks, the difficulties, and the problems I described above. At that point, he caught me off guard with a novel thought: "Sure, but I would call myself a *tribal vegan*," he told me. I did not know what he meant, so I asked him, and he explained it to me with an example: "Normally, I never eat any foods of animal origin, of course, but if my grandma makes me meatballs, I *eat* them because I grew up together with them, both with my grandma and her meatballs. There is a connection that I do not feel like breaking completely." Together, after this conversation, we came to the conclusion that the reason he does this is because he prefers not to make his grandmother suffer by refusing to eat the food that she prepares – "My grandmother grew up in another culture, she would not understand my choice" – instead, he accepts his own "impurity" or moral inconsistency. Of course, every so often, even in this case, there is no fixed rule. If the grandmother invited him to eat meatballs every day, this would create a further problem that the young man would need to address, making sometimes a different move and refusing the meatballs. The logic of caring also includes choices: but these are chosen on the spot, on the road, considering all the variables that may be considered salient. They are not general-based, abstract choices. This is a nice example of dietary caring: a situational diet that corresponds to events. It corresponds: that is, it *responds responsibly and from time to time* to what experience dictates. The logic of caring makes up the rules as we go along, as Wittgenstein would say: you choose, but you choose with every step, depending on your ties, correspondences, tasks, and needs (as when one says, "I had to do so, I had no choice"). It is the logic that guides the space of justice beyond the law ("I did an illegal, but rightful, thing").

Respect and care for the animal is not a formal abstraction. It is not guided by intellectual volition. It takes everything into account, follows the flow, and "decides" by moving, of course, but in a strategic, temporary sense. There is absolutely no proper nutrition. There is not a right taste that is forever. There are no men or animals. We are all animals, and there are people that talk and live with dogs like they were humans: the affections and interests are produced in the intersection field, in the meshwork of forces composed of all the entities with which one lives and deals with. So, are we like a butterfly or a lamb? No, we are morphogenetically different from them; however, we *correspond* and we are woven together. Dietary caring is also an art of encountering, and that is what I refer to in one of my books as *wisdom of taste*.[6] The wisdom of taste is an atmospheric wisdom. It does not need to be explicit; it can also be expressed without speech, without words: for this reason, expressive relationships can also be maintained with animals, without "understanding" their language by comparing it to ours. The animal is like poetry, as Derrida had always insisted: it is untranslatable, with respect to the ontological, differential, and analogical model; but it is always continually and processually in translation, on the other hand, on an allusive, lateral, never-directed level.[7] This process produces nonverbal and unreliable relationships and links that function well, though, as an *ecological feeling of life*. In order to have a respectful relationship with the animal, it is not necessary to study it; it is more useful to live together with it. In this sense, then, the wise man does not possess any ideas – understood as solid and definitive objects, a crystallization of the flow of experience – and has no dietary principles but, for this reason, makes the right decisions and always eats well.

Thus, I propose to live the relationship with food: with as little precepts as possible, less a priori rules, and a greater ability and *disposition* to understand what happens. Dis-ponibility is the opposite of a prefixed position. A position, in fact, is a passage, a transitory time. It is not stable, but labile by definition.

References

Capatti, A. (2016). *Vegetit. Le avanguardie vegetariane in Italia*. Lucca: Cinquesensi.

Cyrulnik, B. (Ed.). (1998). *Si les lions pouvaient parler: essais sur la condition animale*. Paris: Gallimard.

De Fontenay, E. (1998). *Le silence des bêtes. La philosophie à l'épreuve de l'animalité*. Paris: Fayard.

Derrida, J. (2008). *The animal that therefore I am*. Fordham: Fordham University Press.

[6] See Perullo (2016).

[7] See Jullien (2015).

Ingold, T. (2013). Anthropology beyond humanity. *Suomen Antropologi: Journal of the Finnish Anthropological Society, 38*(3), 5–23.
Ingold, T. (2015). *The life of lines*. London: Routledge.
Ingold, T. (2016). On human correspondence. *Journal of the Royal Anthropological Institute, 23*, 9–27. https://doi.org/10.1111/1467-9655.12541.
Jullien, F. (2015). *De l'Etre au Vivre*. Paris: Gallimard.
Kaplan, D. M. (Ed.). (2012). *The philosophy of food*. Berkeley: University of California Press.
Lemke, H. (2008). *The ethics of taste*. Lecture at the international conference food value & taste, University of Aarhus. www.haraldlemke.de/texte/Lemke_Gastrosophy.pdf
Lemke, H. (2016). *Ethik des Essens*. Berlin: De Gruyter.
Perullo, N. (2002). *Bestie e bestioni. Il problema dell'animale in Vico*. Napoli: Guida.
Perullo, N. (2011). Mangerida. Per una gastronomia a venire. In N. Perullo (Ed.), *La scena del senso*. Pisa: ETS.
Perullo, N. (2016). *Taste as experience*. New York: Columbia University Press.
Van Dyke, C. (2015). Manly meat and gendered eating. In A. Chignell, T. Cuneo, & M. C. Halteman (Eds.), *Philosophy comes to dinner. Arguments about the ethics of eating*. New York: Routledge.

Anti-speciesist Rhetoric

Matteo Meschiari

Abstract The various laws protecting animals that were established in Nazi Germany (but for the most part were never put into effect) had, among others, the aim of marking the taxonomic and ontological distance between pure animals and impure subhumans (Jews, homosexuals, the Roma). The attention to and respect for the alpha predator and noble animals was a vertiginous ignoratio elenchi of the concentration camps. With analogous fallacy, today's antihuman and anti-speciesist eco-fascism, which regularly makes use of the *reductio ad Hitlerum* ("meat-eaters = Nazis"), avails itself in an irrational and populist way of the rudimentary *argumentum ad personam* typical of xenophobic and racist propaganda. An extreme case is a well-known PETA campaign against eating meat.

1 Introduction

The various laws protecting animals that were established in Nazi Germany (but for the most part were never put into effect) had, among others, the aim of marking the taxonomic and ontological distance between pure animals and impure subhumans (Jews, homosexuals, the Roma). The attention to and respect for the alpha predator and noble animals was a vertiginous *ignoratio elenchi* of the concentration camps. With analogous fallacy, today's antihuman and anti-speciesist eco-fascism, which regularly makes use of the *reductio ad Hitlerum* ("meat-eaters = Nazis"), avails itself in an irrational and populist way of the rudimentary *argumentum ad personam* typical of xenophobic and racist propaganda. An extreme case is a well-known PETA campaign against eating meat. Matt Prescott, developing a parallelism previously used by I. B. Singer, C. Patterson, and J. M. Coetzee, constructs a photographic diptych in which he places side by side the image of a starving cow and a nude prisoner from a Nazi concentration camp. In other variants Prescott creates a visual parallel between pig carcasses and piles of deportees' bodies or between poultry cages and the interior of a barracks in a Konzentrationslager. Matters of "taste" aside (PETA ended up canceling the campaign and issuing a formal

M. Meschiari (✉)
University of Palermo, Palermo, Italy
e-mail: matteo.meschiari@unipa.it

G. Marrone, D. Mangano (eds.), *Semiotics of Animals in Culture*,
Biosemiotics 17, https://doi.org/10.1007/978-3-319-72992-3_3

apology), the episode lends itself to analysis in order to identify the media strategies of radical anti-speciesism. More specifically, it sheds light on the cultural background that makes it possible for racist, xenophobic, and homophobic fringes to penetrate the animalist movement.

2 Contradiction

Pierre Boule wrote *La Planète des Singes* in 1963. Vercors wrote *Les animaux dénaturés* in 1952. Two basic reflections on anthropopoiesis, long before Hollywood and animalist trends. Over half a century later, in 2016, an anti-bullfighting activist was seen in Arles with a peremptory sign: JE SUIS TORO (I AM BULL). Despite the first impression, perhaps rather comical, what we have here is an eloquent metonymy of the problem, that of an anti-speciesism that contradicts itself, and precisely because it contradicts itself, manages to render to its message efficacious. The first contradiction, the breaking down of taxonomic, ontological, ethical, and juridical barriers between humans and animals, demands a voluntary (or conscious) act of which only the human animal is capable. Primus inter pares, humans are the promoters, the supervisors, the guardians, and the punishers of the new *ordo universalis*; in other words all animals are equal but some are more equal than others. The second contradiction, the human vs. animal dialectic, is an ideological construction of anti-speciesism in the name of the elimination of all dichotomies, but contemporaneously the speciesist vs. anti-speciesist dialectic is an antagonist movement in which are aligned nonhuman animals and reformed human animals on the one hand and on the other human monsters who violate nature and live in ethical-moral sin. The objection could be made that the dialectical plane is not the same, that is, that an egalitarian ontological statute to be affirmed through a political fight is being claimed, but the opposition becomes more radical and the political fight assumes gender connotations. Theorists write books, but ground-level activists make use of eloquent metaphors. Animals are the "slaves," the "victims," the "weak," and empathic emotional thought uses the collective pop imagination of incompleteness and disability as leverage: animals as abused children, institutionalized mental patients, raped women, disabled people who are discriminated against. Difference, therefore, is relied on to advocate for equality. The analogical regime of the ontological model (animals = slaves) is inspired by the model of social inhomogeneity (free vs. slaves).

3 Homogenization

In the thousands of rivulets of animalist reflection, there is also another line of thought: identification with the strong, pure, perfect animal. The Nazi theorem that elevated the noble animal to define a fortiori the contemptible subhuman is a paradigm in act in the most extreme fringes of the movement, but more generally the animism and perspectivist ontology that cross all anti-speciesist thought are laying the groundwork of a new speciesist dialectic, which juxtaposes healthy animality with sick humanity. Putting aside the crypto-fascist deviations, the vigilante and apocalyptic antihumanism, and the resentment of gender extended to the human species, it must be noted that the primary political consequence of being (dialectically) on the side of animals is the relinquishing of principles (too human) of social living. On the one hand, JE SUIS TORO means that I no longer accept human rules so I break them for a cause that I unilaterally deem superior. On the other, it means that the bulls killed in bullfights and the journalists killed at Charlie Hebdo are comparable victims. There is no point in trying to rewrite the meaning of the text with a comeback: that is the message. What comes across, and is meant to come across, is that in the bloodbath there is an ontological homogenization: the same suffering, the same loss, the same violence, and, perhaps, the same moral, ethical, and juridical sanction. Given, however, that human law does not adapt to this logic, it is right to take matters into one's own hands and set fire to the house of a well-known defender of bullfighting, throw nails at spectators in the arena, or toss urine in the face of the president of a bullfight. There is no doubt that these acts are completely devoid of any political depth, that they cannot change things or ideas, and that they bear a striking resemblance to the acts of revenge seen in the early days of *squadrismo*. But JE SUIS TORO can also be interpreted on another level, which allows us to identify a third efficient contradiction. The slogan is not merely an affirmation that is vaguely animist; it is an authentic biological short-circuit that inadvertently evokes the ghost of the Minotaur. Humans are animals like the rest, but they are also the most atrocious of animals (the predator, the "ecological catastrophe," the "most dangerous" of all living beings). Thus on the one hand, morphological details of the teeth and stomach are invoked in order to draw improbable corollaries such as "we used to be vegetarian so we must be vegetarian now too" from our archaic biology; and when this same biology could, on the other hand, highlight a genetic and cognitive distance between the species, it is dropped in the name of an absolute culturalism, which is the mythological construction of the wolf and the lamb. In a society of emotions, it is difficult to separate logic from poignancy. But using emotion to circumvent logical thought is clearly a technique of persuasion; this is precisely what is happening to animalist thought, where theorists keep their hands clean by leaving the real media work to be done by the extreme fringe. And in this downward spiral, not only does the chain of reasoning fall apart, but the premise of a lynch mob-style, metaphysical solution to the problem is also created. Let us examine how.

4 Emotionality

In 2015, the story of the Copenhagen "zoo of horrors," where cases of "animal cruelty" were alleged to have taken place, made headlines around the world. In particular, after Marius the giraffe, a year-old lioness was euthanized and dissected before an audience of children. The press and public opinion, rather than concentrating on and debating the absurdity of a system of captivity in which aggressiveness and overpopulation are in extremis parameters for thinning out zoo guests, expressed extreme horror at the involvement of children in such a "barbaric" act. No matter that the animal was dissected several months after its death, it was nevertheless deemed an act of "cruelty." And no matter that the dissection was performed for educational purposes, the press spoke of "ripping apart" an animal. This (dys)perceptive slip is interesting, this extension of the idea of cruelty and violence from the living being to the cadaver, as if torture and vivisection could continue postmortem, in "another life." Indeed, that is precisely the point: if an anthropologist wanted to write a book entitled *L'animisme aujourd'hui*, a significant portion would be dedicated to animalism and anti-speciesism, and the anthropological perspective helps us to understand that the substratum that makes this belief possible is the new society of feelings, where the phenomenon of anti-speciesist propaganda floats as if in amniotic fluid. Because it is not sufficient to say that we live in a time when rational thought has been overtaken by the blackmailing logic of emotions, we cannot limit ourselves to making light of the millions of mentally unstable people who fill social networking sites with kittens and puppies and express their dissent, any dissent, with a "like."

5 Animism

This double joint of emotional animism and technological endogamy, where homogeneous groups rub shoulders online in a crescendo of exasperated tones, is transforming the western idea of the animal into something that is very different from the previous model. We have gone from a regime of "maximum availability" maintained until recently by the Judeo-Christian culture, to a regime of "absolute intangibility" advocated for by an underground gaseous neo-animism. On the one hand a tough cosmological system, on the other a generic and personal willingness to believe. Despite the existence of some anti-speciesist proselytizers armed with media tools and a desire to dogmatize, for the most part, anti-speciesist animism is equipping itself with pop cults and beliefs, essentially founded on illusions and analogies between *images*. Not a structured credo, but a system of doubts and approximations of thought that does not grasp (or that neglects) the ontological problem of animality and that prefers to find its ideological adhesion (animalism, anti-speciesism, veganism) on mere appearances of meaning. While the development of a doctrine from the top down produces intellectual objects within reach of but a few, from the bottom up, we see the nebulous formation of an intuitive system of representations. And this animism, it must be clarified from the beginning and finds support in the animism of images: humans continually exercise a sort of empathic metacognition on things; in other words we have the capacity to imagine in others (humans, animals, things) the same mental states that we are capable of recognizing in ourselves. This is even truer with images, which more than simple visual objects are zones of possibility in which the projective imagination acts as much pre- as it does postproduction of the image. A real animal – but even a photographed, drawn, or sculpted animal – puts the observer in a contemplative position and urges her/him to establish an intimate relationship with the icon creature. Images are "alive," as W. J. T Mitchell says, but the image of an animal is *even more alive*. Thus there are millions of images of animals that are "so cute!" and "so sad!" dumped onto the Internet, which on the wave of sentimentality produces a shift (semantic and over time cognitive) from information to opinion and then from opinion to belief: news stories are no longer verified, reflected on, or deconstructed with an analytical eye; they simply call for adhesion by a critical assent or dissent.

6 Belief

This shortcut from thinking to feeling means substituting scientific (or scientistic) storytelling with emotional storytelling, with the tale of passions and feelings, through a constellation of images that function as sensitive catalysts in a regime of incomplete, allusive, misinformed belief. And it is precisely in lack, emotion, and imagination that belief finds fertile soil for transforming banality into meaningfulness, doubt into arrogant certitude, private morality into public ethics. It is that mechanism that Carlo Severi defines as "reluctance to not believe": "the connection that is established between the person and the representation (what I have called the psychological aspect of belief) is that distressing/persistent combination from the emotional standpoint and fragile from a logical one, which produces a projection. This connection through doubt has nothing of the serenity of a profession of faith." But what is important is that this "I don't know if" is based, at least in our case, on *Homo sapiens*' natural predisposition toward animism: the "weakness to admit a few isolated exceptions" is possible precisely because our species needs to ascribe a soul to almost everything, to dig paths of meaning in its existential exposure to doubt, a soul to animals and a soul to images of animals as well. This is nothing new, of course. Facebook is filled like Lascaux with icons that help us to think about/feel some aspect of the world that is important to us, but with the difference that in the absence of a cosmology, the problem is reduced to a particular frame of mind in which it is the anguish of incertitude (both individual and collective) that generates projections. I am not referring to the apocalyptic antihumanism of radical anti-speciesists, or to the increasingly viral equating of carnivores to pedophiles, or to the pack-like aggressiveness toward the terminally ill who commit the unforgivable sin of taking pharmaceutical drugs tested on animals. Rather, I am thinking of captions such as "I kiss pure souls worthy of unconditional love" which accompany pictures of puppies and pets and which help us to understand what kind of animism contemporary images of animals is flattened upon. Easy love, speechless and generally devoid of Gregory Bateson's double bind, relegates animals to an emotional function that tends to provide a satisfying counterbalance to the difficulties and the lack of affection that characterize a solitary life and the absence of community. It is therefore understandable that Marius the giraffe, Daniza the bear, and Cecil the lion or the lioness dissected in Copenhagen are the beneficiaries of not only anti-speciesist empathy but also, perhaps (and especially), of an anthropomorphous transfer. Perhaps on that anatomy table there is the subjective shame of the nuder nudity laid bear before an audience of children? Otherwise why the indignant disgust at the idea of a child-spectator? A bit of anthropological knowledge would suffice to resolve the impasse: animal anatomy fascinates children who have not yet been taught to be disgusted by it, because the whole of human and animal organs was the first complex inferential system that our species practiced on. That is where organicism and structuralism come from, and a lesson in animal anatomy offered to a child is simply a cognitive gift, not the sensationalizing of an act of cruelty perpetrated postmortem on an innocent soul. The tendency, however, is toward an animism that we could define as Disneyfied: a world of animals perceived as children and of children perceived as baby animals.

7 Strategy

In an attempt to make its point, anti-speciesist rhetoric is based on a few ideological and cultural premises that simultaneously act as media strategies:

(a) EXCLUSIVENESS: radical anti-speciesism is adopting the means and tenor of political and religious monologue. Profoundly differentiated within, its solidity derives from the universal dialectic of "us" vs. "them." A language (and more and more frequently, acts) of violent conflict intercepts a generic and viscous social malaise that thrives on tragic tones ("either with us or against us"). Verbal micro-fascism and hit squads online are replacing civil dialogue with the unilateral tones of accusation, vigilantism, fanaticism, and intimidation and the herd mentality. This violent construct, with vindictive and at times apocalyptic overtones, highlights the "chosen-damned" antinomy outlined above. Only the "just" will be saved. All others will perish in the flames of hell.

(b) IRRATIONALISM: despite the fact that anti-speciesism looks to science for support (evolution, neuroscience, physiology), its discourse remains essentially an ethical-moral narrative. Being a tale, its effectiveness does not come from its degree of veracity or rationality, but from its capacity to transform the spectator into a character. In order to achieve this identity transformation, it has to make the spectator suspend rational thought, in two ways: by looking for an emotional chink and by using fallacious logic. Emotion and discursive fallacy open the door to mystical thought and spiritualism. Because of this, anti-speciesism is permeable to alternative philosophies and subcultures (new age, theosophy, conspiracy theories, shamanism, millennialism, pop animism, etc.).

(c) FIDEISM: despite the meticulous pseudo-rational arguments, when faced with the impossibility of providing decisive proof of rationally untenable hypotheses ("the suffering of a mosquito and that of a calf are identical"), a final act of faith is required of the anti-speciesist. This happens either through emotional adhesion ("I feel this is right") or through reasonable doubt ("I am not certain, but it may be true"). In both cases the anti-speciesist discourse is impelling only to a certain point, after which it is the individual conscience that intervenes to seal the deal. This shift from logic to faith does not happen once and for all, but rather becomes a recurring way of viewing and solving problems, including those that arise when confronted with the ideas of others.

(d) POPULISM: anti-speciesism has armed itself with a propaganda apparatus in which the enemy is represented as a monster that is assaulting the perfection and purity of animals ("meat-eaters are Nazis/barbarians/serial killers/pedophiles"). This propaganda produces not only rhetorical statements but also encourages cognitive habits. Critical interpretation of phenomena is replaced with emotional involvement; rather than analyzing the problem, the focus turns toward empathy. Indignation, tenderness, pity, pride, and anger are the basic grammar of feelings that guide the handling of information. Enormously amplified by the social network platform, the system of emotional blackmail replaces careful interpretation with hasty opinion.

(e) ANTHROPOCENTRISM: despite the fact that the main goal of anti-speciesism is to dismantle the ontological and social primacy of humans, it is nevertheless humans who must guide the advocated transition from a society of exploitation to a society that is more just for animals. Despite the fact that this transition must come about through greater and equal rights for animals, it is still humans who must guarantee this new ethical-juridical system. On the one hand, antihumanism and post-humanism are spoken of; on the other hand, the enlightened and paternal human (some, not all) is an indispensable ally-collaborator. From this emerges the other face of anthropocentrism: apocalyptic nihilism ("no more humans, only animals").

8 Discussion

A radical reframing of the human-animal relationship is no doubt necessary, especially in a regime of fanatical consumerism. The global food industry, intensive exploitation, and unspoken suffering are all real and urgent problems. A true reconsideration must be based on a broader awareness, but anti-speciesist ideas and practices tend to radicalize the terms of the problem and tactically polarize the positions of the social actors. The reason for this is that the practices (right or wrong) rework ideas (right or wrong) based on the logic of mass communication. In particular, the strength (media) and the weakness (political) of anti-speciesism respond to that spectacular system identified by Guy Debord, even (and especially) in counterculture movements. Tearing down the ontological barriers between humans and animals on this terrain means heading toward two complementary scenarios with which humanity has been well acquainted since the Neolithic: isolating a caste that makes political decisions in the name of an ideal that supersedes the needs and rights of the individual, defining a new operative taxonomy that separates good animals who make decisions (anti-speciesists) from bad animals worthy of being punished, reeducated, and suppressed (speciesists). The radical right's flagrant infiltration of the anti-speciesist movement should alarm everyone. Rarely, however, do "left-leaning" anti-speciesists make the effort to distance themselves because, for the cause, the ends justify the means. Thus, between "empathic feud" and "supermodern neo-animism," the dialectic "pure animal" vs. "impure carnivore" is not only part of a strategy of ethical propaganda, but becomes the ideological basis for new scenarios of social exclusion.

References

Accarino, B. (2013). *Zoologia politica. Favole, mostri e macchine*. Milano: Mimesis.

Agamben, G. (2001). *L'aperto. L'uomo e l'animale*. Torino: Bollati Boringhieri.

Biehl, J., & Staudenmaier, P. (1995). *Ecofascism. Lessons from the German experience*. Oakland: AK Press.

Bird-David, N. (1999). Animism revisited. Personhood, environment, and relational epistemology. *Current Anthropology, 40*(Special Issue, Culture. A Second Chance?), 67–91.

Brightman, M., Grotti, V. E., & Ulturgasheva, O. (Eds.). (2014). *Animism in rainforest and tundra. Personhood, animals, plants and things in contemporary Amazonia and Siberia*. Oxford: Berghahn Books.

Buscemi, F. (2016). Edible lies. How nazi propaganda represented meat to demonise the Jews. *Media, War & Conflict*. https://doi.org/10.1177/1750635215618619 (on line).

Carmichael, R. (2002). *Becoming vegetarian and vegan. Rhetoric, ambivalence and repression in self-narrative*. PhD Dissertation, Loughborough University.

Cimatti, F. (2013). *Filosofia dell'animalità*. Roma: Laterza.

Debord, G. (1967). *La societé du spectacle*. Paris: Buchet-Chastel.

Descola, P. (2005). *Par-delà nature et culture*. Paris: Gallimard.

Despret, V. (2008). The becomings of subjectivity in animal worlds. *Subjectivity, 23*, 123–139.

Frye, J., & Bruner, M. (2012). *The rhetoric of food. Discourse, materiality, and power*. London: Routledge.

Ginzburg, C. (1966). *I benandanti. Ricerche sulla stregoneria e sui culti agrari tra Cinquecento e Seicento*. Torino: Einaudi.

Ginzburg, C. (1986). Spie. Radici di un paradigma indiziario. In *Miti, emblemi, spie. Morfologia e storia*. Torino: Einaudi.

Ginzburg, C. (1989). *Storia notturna. Una decifrazione del sabba*. Torino: Einaudi.

Graham, H. (Ed.). (2014). *The handbook of contemporary animism*. London: Routledge.

Hallowell, I. A. (1976). Ojibwa ontology, behavior and world view. In I. A. Hallowell (Ed.), *Contribution to anthropology* (Selected papers of A. Irving Hallowell, pp. 357–390). Chicago: University of Chicago Press.

Ingold, T. (2000). Essays in livelihood, dwelling and skill. In *The perception of the environment*. London: Routledge.

Kohn, E. (2013). *How forests think. Toward an anthropology beyond the human*. Berkeley: University of California Press.

Kozinets, R. V. (2010). *Netnography. Doing ethnographic research online*. London: Sage.

Laughlin, C. D. (2011). *Communing with the gods. Consciousness, culture, and the dreaming brain*. Brisbane: Daily Grail Publishing.

Marrone, G. (2011). *Addio alla Natura*. Torino: Einaudi.

Maurer, D., & Sobal, J. (Eds.). (1995). *Eating agendas. Food and nutrition as social problems.* Piscataway: Transaction Publishers.

Mitchell, W. J. T. (2005). *What do pictures want? The lives and loves of images.* Chicago: University of Chicago Press.

Pinotti, A., & Somaini, A. (Eds.). (2009). *Teoria dell'immagine. Il dibattito contemporaneo.* Milano: Raffaello Cortina.

Safran Foer, J. (2009). *Eating animals.* New York: Little, Brown and Company.

Santos-Granero, F. (Ed.). (2009). *The occult life of things. Native Amazonian theories of materiality and personhood.* Tucson: University of Arizona Press.

Severi, C. (2004). *Il percorso e la voce. Un'antropologia della memoria.* Torino: Einaudi.

Shepard, P. (2011a). *The tender carnivore and the sacred game.* Athens: The University of Georgia Press.

Shepard, P. (2011b). *Thinking animals. Animals and the development of human intelligence.* Athens: The University of Georgia Press.

Singer, P. (1975). *Animal liberation. A new ethics for our treatment of animals.* New York: Random House.

Tarantino, C., & Straniero, A. (2014). *La bella e la bestia. Il tipo umano nell'antropologia liberale.* Milano: Mimesis.

Underberg, N. M., & Zorn, E. (2013). *Digital ethnography. Antropology, narrative, and new media.* Austin: University of Texas Press.

Viveiros de Castro, E. (1998). Cosmological deixis and amerindian perspectivism. *Journal of the Royal Anthropological Institute, 4*, 469–488.

Wunenburger, J.-J. (1997). *Philosophie des images.* Paris: PUF.

Aesthetics of Nutrition, Ethics of Animality: The Packaging of Vegan and Vegetarian Products in the Italian Organised Distribution Market

Ilaria Ventura Bordenca

Abstract The topic of this paper concerns the packaging of vegan and vegetarian products in the Italian organised distribution market. Large-scale organised distribution and vegetarianism's choice of packaging is related to current dietary trends which, as is often the case with eating choices and habits, intertwine as they massively spread. Which are the visual and verbal strategies that are adopted to give meaning, in the Italian culinary landscape, to vegan and vegetarian foods sold through large-scale distribution? The aim is to outline, through the methodology of semiotics, the sensorial mechanisms and cultural grids that shape this food universe and the way it interacts with other categories of foods: specifically, with the traditional omnivore diet, on the one hand, and with the so-called biological nutrition, on the other. Here we will investigate a specific area of *veg* food culture or rather the supermarket product, ranging from preserved foods to pre-cooked condiments, from frozen main courses to side dishes ready in 5 min.

1 Introduction

If abstinence from eating meat is in principle the just and inevitable passage to consistently protect the welfare of animals, and is therefore the expression of the ethical-political position of animal lovers, it is true that vegan and vegetarian foods are increasingly becoming part of variable healthy food styles and not always linked to animal welfare ethics. Eggs, cheeses, meat, and cured meat products are foods that in our times are perceived by some as enemies when eating, as antagonists of a system of nutritional values subject to an equally organised block of sociocultural values within which *fat* – with its double meaning of substance contained in foods and of external effect on the body – is something to avoid, to escape from and, in some cases, to stigmatise (Niola 2015; Ventura Bordenca 2014a). On TV, diet

I. Ventura Bordenca (✉)
University of Palermo, Palermo, Italy
e-mail: ilaria.venturabordenca@unipa.it

G. Marrone, D. Mangano (eds.), *Semiotics of Animals in Culture*,
Biosemiotics 17, https://doi.org/10.1007/978-3-319-72992-3_4

reality shows on obesity and extreme slimming do nothing but remind us of this. So, we choose to eat vegetarian for a supposedly greater lightness and digestibility, because these are low-calorie foods and they don't make us fat: here the aesthetic reasons emerge.

The intertwining of ethical, aesthetic, and hygienic motivations ends up reinforcing the positive value of vegetable foods in our society, hence also increasing its commercial and brand value. From this, the ever-widespread presence of vegan and vegetarian products in the large-scale distribution market, not only in specialised and niche stores: the massification of the ethically correct diet is happening.

Lastly, the choice of working on the packaging is derived from the role that this communication support holds in communicating at the point of sale: not simply a containing box, a container of a content already loaded with meaning, but an important communicative text through which the identity of the product is created, making it recognisable, different, together with the brand, from other brands through shapes, colours, words, images, textures, and materials. Packagings are complex texts that also have the task of giving instructions on the product and its use: a list of ingredients (today a very important cognitive phase of the so-called conscious consumer), usage guidelines, recipes, and ways and times of cooking and preservation but also best before dates, prohibitions, warnings, and so on (Ventura Bordenca 2014b). Even *veg* food cannot do without its packaging to provide information and position itself in the arena of consumption.

2 Tastes to Forget

In *Animal Liberation* (1975), a text that is considered one of the manifestos of animal welfare ethics, author Peter Singer, after addressing the philosophical grounds of animal rights, discusses the need for the transition to the vegetable diet as one of the changes to face to follow animal welfare ethics coherently and on a daily basis. After debating on the suffering of animals through the crude and detailed description of the methods of breeding for food and animal experimentation practices, Singer dedicates many pages to the concrete way in which to switch from the omnivorous diet to nutrition without meat and fish, first (vegetarianism), and without any animal-derived product, after (veganism).

The book, as a *manifesto*, is the declaration of a system of values around which we recognise ourselves and establish a new collective identity, but also by presupposing the attestation of the opposite system of values, that of the anti-subject to be fought: speciesism, with respect to which the choice to abstain from animal meat is a means of battle. In addition, the setting of Singer's text presupposes an enunciatee to involve – more precisely, to upset – to cause an actual transformation: stop eating meat and fish. Knowing, seeing, and reading "what happened to your dinner when it was still an animal" (this is the title of the chapter about slaughtering) would involve emotional upheaval, the revelation of the truth, and the power of concrete change. *Informing* is an obligation of the animalists; *knowing* is a duty of ordinary people who still eat meat; giving it up is to *put into practice* what comes from it. A passionate-cognitive strategy that is very common in social communication campaigns, even those about animal welfare which very often use gory images and strong reports precisely for obtaining, at least in their goals, an emotional, mental, and pragmatic transformation of the enunciatee.

In Singer's book, in fact, there are several changes of strategies: from the initial philosophical and historical argument, to the gritty accounts of torture carried out by animal husbandry and medical research on animals, to the final advice on how to become vegans without too much effort. It is with some astonishment that the reader, after pages of a realism at times disturbing, faces a quasi-cookbook in which Singer suggests ways, times, ingredients, and dishes to stop being carnivores and to happily embrace the vegetable diet. The author perfectly understands the objections and difficulties that the abandonment of the omnivore diet implies, particularly regarding matters of taste and nutrition, and anticipates the most common resistance and the most common prejudices on vegetarian diets: What to eat instead of veal? Will I have to give up fish too? What does tofu taste like? How do I cook it? And above all, will I keep strong without meat and fish? Where will I get protein from?

The author suggests a gradual abandonment of meat to reach veganism, which is the ultimate fulfilment for boycotting the system, and of animal liberation. And here, Singer once again understands the difficulty, in our Western food culture, of eating egg-free and dairy-free products ("Don't feel obliged to make tremendous efforts to eliminate any food that contains dairy products. Eliminating speciesism from your own eating habits is a very difficult thing to do all at once", p. 208 Ita. trans.). A sort of non-traumatic adjustment for the potential vegetarian reader.

As for vegetarian cooking, Singer writes, it is rich, interesting, and full of unusual foods coming from China, Europe, and the Middle East, far from the monotony of the Western diet based as usual on meat (in particular referring to the Anglo-Saxon one). The common dish on our tables consisting of a meat and vegetables combination, once abandoned, leaves room for fantasy and new flavours: "millet, whole grain or buckwheat can be the basis of a pleasantly different dish"; "Tofu dumplings to season spaghetti"; "Italian vegetarian lasagne with a little salad"; "Whole grains and vegetables"; "good Chinese dishes are wonderful combinations of one or more high-protein ingredients with lightly cooked vegetables and rice"; and so on with his culinary suggestions.

As in any dietary system change, that is, of eating habits, there's the question of taste which, in addition to dealing with the aesthetic and perceptive experience, is related to custom, tradition, individual preferences, family recipes, festivities, and social rituals.

In an autobiographical essay-novel dealing with the conversion to vegetarianism of the author, *Eating Animals*, Jonathan Safran Foer (2011) addresses exactly the problem of tastes to forget. There's the awareness that abandoning certain dishes implies both a strong individual willpower and the prospect of rebuilding new habits and new family rituals around other foods:

> Giving up the taste of sushi or roast chicken is a loss that goes beyond renouncing to a pleasant gastronomic experience. Changing our diet and deleting certain flavours from our memory is kind of a cultural loss, something that is forgotten. But perhaps it's worth

> accepting, if not even cultivating, this forgetfulness. To remind myself of animals and how important their welfare is to me perhaps I should let go of some tastes and find other pretexts for the memories they once helped me to preserve. (Page 210 It. Trans.)

A complex mechanism of *cultivating oblivion* (forget what we used to like), of *reconstructing memory* (remember that the animals we have on our plate once were living beings), and of *building up future memories* (creating new eating habits). This is the case of the turkey at Thanksgiving. Safran Foer, who is about to prepare his first turkey-free Thanksgiving dinner, raises the opportunity to create a new ritual of gratitude and different holiday memories for his children, no less dear and significant than his childhood ones around the turkey:

> And what if there was no turkey ever again? Would it break, or would it harm, the tradition, if we skipped it and fed ourselves with stewed sweet potatoes, homemade sandwiches, string beans with almonds, red cranberry compote, potatoes, butter pate and pecan nuts? [...] It's not such an absurd thought. Your loved ones around the table. Listen to the sounds, smell the aromas. There is no turkey. Is the party perhaps ruined? Is Thanksgiving no longer Thanksgiving? (P.269 It. Trans)

The ritual, and therefore culturalised, dimension of foods is revealed and is even stronger when it comes to abandoning and replacing them.

3 Substitutions

Lots of *veg* products are based on exotic, mainly Asian, ingredients such as tofu and tempeh or created ad hoc specifically for vegetarian and vegan cuisine to replace meat and other animal-origin foods, such as "wheat's muscle". All in all, these are little-known foods in our culinary culture.

Looking at some culinary blogs, the evident reference to meat and cheeses is constantly found: tempeh is also known as "soya meat", tofu as "soya cheese", Mopur is called "vegetable meat", and "wheat's muscle" is obviously named as such just to remind the appearance of meat. The proposed recipes are also re-elaborations of dishes that normally have meat as main ingredient: roasts, stews, burgers, etc.

What is implied by such a type of substitute product description? On the one hand, that the object of the communication needs an introduction, being contextualised in a family setting, and in parallel that the enunciatee is to be educated, introduced, and accompanied in a new world of food.

At the basis of these discussions, there is a *substitution* process, a phenomenon that is common to almost all diet phases. Substitutions, however, are not all the same and provide changes in the whole system involved. During a slimming diet, for example, not only some foods considered hypercaloric should be replaced with others but it is widespread practice that for each substitution the person following the diet usually has a range of meal alternatives, equivalent to each other. If from a certain point of view dietary foods are replaceable because they have the same value (e.g. calories), it surely cannot be said that they are in other aspects: flavour, personal preferences, techniques and cooking times, etc. From here the relapses and difficulties, for those who undertake a diet, to get used to the new type of nutrition and stick to it.

Diet is not just a change in food intake; it has consequences on all of the subject's values: aesthetics (tastes and distastes), pragmatics (culinary habits, rituals), affections (traditions and preferences),and social life (diet management within family, friends, working environments).

Furthermore, product substitutes for vegetable diets, as mentioned previously, are not well-known in the Italian market, in terms of taste, nutritional properties, cooking techniques, possible recipe creations, combinations, and so on. It poses a problem that is of a communicative nature and therefore cultural: How to make this product known and how to explain its use? How to inform the multitude of consumers who are increasingly buying – and eating – far from rigid dietary classifications but balancing vices and virtues at the table? And how are the issues of taste and nutrition tackled?

4 Trusting Imitation Products

On supermarket shelves, vegan and vegetarian food packaging does not differ from that of products for omnivores, either in shape, colours, or materials used: standard containers, plastic, transparent film, plenty of colours, and great use of reference photo images.

Unlike the packaging of organic products, where there is a rhetoric of reduction (Marrone 2011; Ventura Bordenca 2012) that is necessary to communicate a visual and value-related contrast to the logic of branding and industrialised food production, these packets, more or less, are similar to those of traditional products. By reducing and simplifying the packaging's communication, the organic product strongly differs from the industrial and traditional ones: apparently homely containers, less colours, no plastics, no transparent film, and use of "natural" materials and colours that figuratively recall various images of nature that we often see in our culture. Clearly, this is not an effective reduction of the communicative function of the packaging, but of a sensorial effect: it's the staging of packaging impoverishment through very attentive and planned graphics and design, on the packaging itself. The product in some cases also corresponds to the same logic of detachment from traditional food with the display of product imperfections that emerge as a sign of sincerity (crumbled, small, imperfect products, etc.). The packaging of *veg* products, on the other hand, adopts a strategy of visual imitation of the traditional product.

Also, regarding the names of the food preparations on the market, these come directly from the culinary culture of meat: seitan steak, seitan fillet, soya burgers, soya meat-like strips, soya Bolognese sauce, wheat muscle bresaola, and so on. In this way, a steak, even if it is not meat, remains a steak in the *sense* of steak, an eating style, a type of dish: a main meal, accompanied by a side dish, with wine, to bite in a certain way, eat at certain times, etc. A steak, a fillet, and a burger, from *substances*, in a semiotic sense, of the animal-origin food world, become *culinary forms*.

It seems that a first conclusion that can be drawn may be that the world of *veg* foods needs, so to speak, animal-origin foods to communicate with the consumer and give a sense of edibility.

For what concerns the visual aspect of the packet and in particular the images used to represent the product, a first distinction should be made between reference image packets (usually a photograph with soya burgers, seitan cutlets, etc.) and packets without a reference image: two different enunciatees, one that needs to be educated and the other who already knows how to use, for example, tempeh.

Let's focus on those with the image: What image? How is the product's use being communicated? And its taste? And the famous nutritional values? As for the latter, these are constantly reiterated, because in our culture we are accustomed to associate meat with nutrition. Books, manuals, scientific papers, and other texts circulating in the media supporting the vegetarian diet list substitute products that provide the much-needed protein, trying to disprove the widespread opinion that it can be found only in meat and fish. Vegan cookbooks, for example, very often contain a pre-recipe introduction in which, besides explaining how to replace ingredients, mixtures, and basic sauces for "veganising" a recipe, also give advice on how to ingest the right amount of protein and vitamins. Vegetarianism pays the consequences of contemporary nutritionism and in particular the search for protein (Pollan 2009).

Regarding taste and flavour, here the *camouflage logic* completely prevails because the photographed foods are an imitation of the shapes, colours, and textures of animal foods and their varieties: the cutlet is recognised by the breading, stir-fries and stews from their cutting shape, a steak from its size and glossiness, cured meats are distinguished by their colour and the presence or absence of chromatic variations on their surface (salami, mortadella, bresaola, etc.), and so on (Figs. 1, 2, and 3). The same logic applies to vegetable cheeses, sauces, and fish on sale.

This staging of the distinctive expression elements of various animal-origin foods is a semiotic process that identifies this or that product, which in turn means this or that way of eating. Ironically, however, the emphasis is placed on the distinctive and pertinent aesthetic characteristics of banned foods, meat, fish, and cheeses. This camouflage strategy implies a series of tactics, of visual and verbal means, like the naming and expressive imitation procedures we have seen, with which the *veg* dish is constructed.

To this end, it is interesting to observe another aspect that emerges as pertinent. At pre-culinary stage, many vegan foods have more or less the same colour, from white to brown, while at the stage of consumption, in some cases, they are almost all round "medallions" that are indistinguishable between seitan, tempeh, tofu, etc. unless it is stated on the package itself (Figs. 4 and 5). Here there is a problem of discretisation of these mixtures, of telling about their culinary possibilities, of translating the tastes. The first technique is always the lexical one: to call, that is, the

Fig. 1 "Wheat's muscle" vegan stew

Fig. 2 Vegan sausage

Fig. 3 Vegan mozzarella

Figs. 4 and 5 Lupine seed steak and tofu and seitan medallions

same world objects with different names (burgers, medallions, steaks, etc.) as each of these foods recalls a different way of arranging the plate and eating. The second technique is to distinguish main course and side dish: it is from the union of a soya cutlet and a salad that the dish is constructed, that is, a culturally understood sensorial union. A well-known and comprehensible syntagma appears again: meat-salad, meat-potato, or cured meat and bread, and so on.

These techniques of food taste visual translation, that is, a plastic reproduction of meat, fish, and cheese flavours, and of figurative composition of traditional omnivorous dishes, are explicit food-design operations based on *camouflage*.

As Fabbri (2008) argues, the camouflage aesthetics should not be considered from the truth/lie criteria point of view as compared to an external referent of which the camouflaged object is a copy. On the surface, in fact, meat imitation techniques could be negatively seen, derided, diminished, and criticised as somewhat ridiculous and grotesque copies of something that is original and therefore of greater value. According to Fabbri, in the semiotic function of camouflage, the relevant factor is *effectiveness*: we are not within a truth/lie system that would carry the opinions just expressed but in a system of belief. The enunciatee is knowingly buying imitative foods within a precise value-related contract, in a belief system with known rules, where imitation itself is no longer relevant – they are all imitative – and importance is given to how much that food can really replace meat, fish, or cheese from a visual (it really looks like a steak), taste-related (it's satisfying, I like it), and nutritional (I am full, I feel strong) point of view.

5 Aesthetics of Nutrition, Ethics of Animality

The omnivorous culinary imagery serves as a model for vegetarianism, at least in the cultural sphere we have analysed: both in the sense of a *figurative universe* to draw from so to capture images and taste perceptions and in the sense of a *social discourse* from which to repropose culinary habits, ways to cook and eat food. This is despite the fact that *veg* diet offers food that is totally different, a mixture of foods that has little to do, either at material or philosophical level, with the omnivorous diet. Yet, the dialogue between the two worlds is constant and necessary from a communication point of view.

If we look again at the comparison with the organic sector, we will notice that the design of *veg* and *organic* food packaging reflects two completely different aesthetics to which two different logics of connection to traditional food (with the term "traditional" we are aware of subsuming and normalising two different views: omnivorous compared to the vegan world and industrialised compared to the organic world) correspond. On the one hand (veg vs. traditional), there is a *logic of imitation* and on the other (organic vs. traditional) a *logic of originality*. This opposition, from a terminological point of view, does not involve judgments, for which the imitation would produce copies and therefore should be judged negatively, whereas originality always has a positive value of authenticity. We are not talking about specific issues of creativity or authorship; we are far from the artistic field.

There are two general ways of producing texts within a certain culture. In the case of the *veg* industry's commercial communication, as we have seen, we are faced with the reproposition of well-known and familiar styles and expressive features. Reproposition is never a perfect mimesis, and changing the enunciative context creates the effects of originality due to the rehandling and resemanticisation that inevitably results from it (vegetable meat and fish are not the meat and the fish of the omnivorous meals, but something different that did not exist before and which is among the foundations of a specific food philosophy).

In the case of the *organic* sector's commercial communication, we are faced with a strong detachment from the most popular expressive features that usually present and promote food. However, as we have seen elsewhere (Ventura Bordenca 2012), the *organic* communication to differentiate brands and products, and to satisfy its marketing goals, draws on the various images of the idea of nature that already exist in our culture, in order for various forms of natural character, more or less stereotyped, to be reproposed and revived (rustic, exotic, 0 km, scientific-medical, etc.). There are, therefore, echoes and references that create imitative effects between pre-existing imagery.

We can now refer to Lotman (1972), who speaks of two opposite aesthetics with which to design and distribute products within a culture: *aesthetics of identification* and *aesthetics of contrast*. The *aesthetics of identification* "are based on the full identification of life phenomena through the standard models already known to the public and falling within the system of rules" (Lotman 1972, p. 339 It. trans.). This is based on the simulation of existing standards and rules. The aesthetic of contrast occurs when the text "counterposes its original solution to reality simulation methods which the reader is accustomed to" (p. 341 It. trans.).

In particular, Lotman refers to the creation of artistic texts, but his model implies a general mechanism that can be applied to all types of texts as sociocultural objects, thus also to food and the communication around it. It is neither a matter of copying nor inventing ex novo but of opposition to the audience's knowledge and expectations, on the one hand, and society's production standards, on the other.

Therefore, the large-scale distribution's sphere of *veg* food offers identification aesthetics and exploits the pre-existing taste-related visual and culinary universe to make it its own and gain space in the arena of gastronomic media. The *organic* food sphere, on the other hand, is based on contrast aesthetics, with food and packaging design characteristics opposite to those of the traditional sector.

In either case, the result is the construction of two dietetic discourses, both nutritional and ethical, which aim at being unconventional, of opposition and of change. And for this reason, they cannot but refer to the whole world of omnivorous and/or industrialised consumption from which they derive and dissociate with. Two opposite visual strategies – at least in the current historical and economic context – which, however, correspond to a common countertrend, the same instance of dissent and of new ethical proposition.

Since these are visual and value-related identities that are produced by difference, we can refer to that anthropological mechanism recognised by Lévi-Strauss for tribal masks (1985) and that can be found in various contemporary brand positioning phenomena (Floch 1997). This is an inversion between expression and content level between enunciative instances in some way competing: when two groups or subjects are characterised by opposite expressions, the content of their message will be almost the same; on the contrary, when they share the same distinctive signs, the content will be the opposite.

In fact, *veg* and *organic* communications are two *gastro-spheres* with very different aesthetics, characterised by opposite visual systems, but at value-related level, they share the same will to change as compared to normal, so to speak, production logics and food consumption.

Table 1 The different aesthetics of veg and organic foods

	Veg packaging	Organic packaging
E	Imitation	Originality
	Aesthetic of identification	Aesthetic of contrast
C	Ethical food	

Table 2 Common expressive characteristics conveying opposite content in veg and traditional packaging

	Veg packaging	Traditional packaging
E	Rhetoric of "normal" and industrialised food	
C	Inedibility of animal meats	Edibility of animal meats

Table 3 Opposite expressive characteristics conveying the same content in organic and traditional packaging

	Organic packaging	Traditional packaging
E	Rhetoric of reduction and imperfection	Rhetoric of abundance and perfection
C	Edibility of animal meats	

Such value-related instance is nothing more than a clear and explicit *nature policy*, a stance with respect to the limits and possibilities of the human on the non-human, on the boundaries of the communities we consider as social (Latour 2000): how and how much to consume nature, which processes to consider acceptable or not, and which and what kind of non-human actors to include in the collective and thus protect (Table 1).

By entering into the details of the two philosophies, however, things become more complex. Regarding the issue of non-human actors, we are confronted with two completely different ethics of animality: in the *veg* world, we abstain from meat because animals are untouchable; in the *organic* world, animals are possibly respected and treated well, raised according to certain criteria, but eventually eaten and savoured. That is to say, with respect to the category of the edibility of meat and its derivatives, the two diets are in opposite positions and in principle communicate an opposite message (although let's not forget the existence of all intertwining forms of organic vegetarian).

And in fact, with regard to the aesthetics of foods for omnivores, things change and split in two. As we have seen, *veg* foods share the communicative methods of food for omnivores, for both the appearance of the food and the packaging design. To the same expressive characteristics, the opposite content corresponds: the prohibition to eat meat and its derivatives, of which those foods are only imitations. *Organic* foods, instead, have expressive characteristics other than those of common and popular food but share with them the value of animal meat edibility (Tables 2 and 3).

To sum up, if we look at the *veg* and *organic* worlds and make a comparison, we will find that they have different expressions but same content with respect to a general message which is more or less revolutionary and of ethical change. If we look at them separately from another point of view, compared to the traditional omnivorous food, we will find a different situation, that is, respectively, same expression but different content (veg vs. traditional) and different expression but same content (organic vs. traditional). The different dietetics around our society, therefore, are doing nothing more than talking to one another, having a more or less explicit confrontation, incorporating elements of each other, and strengthening each one's own identity.

References

Fabbri, P. (2008). Lo sguardo dell'altro. Strategie del camouflage. E/C, online magazine of the Associazione Italiana di Studi Semiotici, www.ec-aiss.it
Floch, J.-M. (1997). *Identità visive*. Milano: Franco Angeli.
Latour, B. (2000). *Politiche della natura*. Milano: Raffaello Cortina.
Lévi-Strauss, C. (1985). *La via delle maschere*. Torino: Einaudi.
Lotman, J. M. (1972). *La struttura del testo poetico*. Milano: Ugo Mursia Editore.
Marrone, G. (2011). *Addio alla Natura*. Torino: Einaudi.
Niola, M. (2015). *Homo Dieteticus. Viaggio nelle tribù alimentari*. Bologna: Il Mulino.
Pollan, M. (2009). *In difesa del cibo*. Milano: Adelphi.
Safran Foer, J. (2011). *Se niente importa. Perché mangiamo gli animali?* Parma: Guanda Editore.
Singer, P. (1975). *Liberazione animale*. Milano: Il Saggiatore.
Ventura Bordenca, I. (2012). Nature in vendita. Il packaging dei prodotti biologici. In G. Marrone (Ed.), *Semiotica della natura*. Milano: Mimesis.
Ventura Bordenca, I. (2014a). Questa non è una dieta. Linguaggi e strategie del dimagrimento. In G. Marrone (Ed.), *Buono da pensare. Cultura e Comunicazione del Gusto*. Roma: Carocci.
Ventura Bordenca, I. (2014b). *Che cos'è il packaging*. Roma: Carocci.

Forms of Animality: The Dog

Dario Mangano

Abstract Animals are not the object of zoosemiotics as discussed here, neither is their language nor any other system of signs that affects them directly as living creatures that communicate, but the *forms of animality*, forms through which the concept of animal itself is socially constructed and transformed over time. Now, if there is one place in which such forms are plain to see, it is in those areas that focus on their nutrition. Looking at feeding animals as a semiotic activity, this paper wants to focus on what I call a *nutritional pact*, an implicit agreement that binds humans and animals, and that is being continuously negotiated by cultures. A relationship that advertising deeply contributes to build.

1 An Idea of Zoosemiotics

For a science that considers the nature and values of entities as subordinate to the relationships they assume within a structure, a consideration of the animal must therefore entail a consideration of the human being. This is not, however, a moral question: the amount of beast in humans and the amount of humanity in animals are not our concern, at least not directly. The aim is not to establish priorities nor to find similarities or differences to the animal kingdom that may be revealed by a semiotic approach. What interests us are the mechanisms by which human societies view the animal as other, whilst simultaneously producing an idea of themselves. Not where the frontier, ontologically speaking, between the two lies, but how it is constructed and what is done from time to time to change its features. Semiotics has, over time, moved from a static perspective (what does it signify?) to a dynamic one (how does it signify?), in which the investigation's objective is no longer the signified, but the meaning, and the model of reference is no longer the sign, but the text (Marrone 2010).

We know very well what animals – and dogs, in particular – signify in contemporary society. And if the sheer number of four-legged friends that populate our cities wasn't enough, we have the books, photography, conversations and, of course,

D. Mangano (✉)
University of Palermo, Palermo, Italy
e-mail: dario.mangano@unipa.it

G. Marrone, D. Mangano (eds.), *Semiotics of Animals in Culture*,
Biosemiotics 17, https://doi.org/10.1007/978-3-319-72992-3_5

specialist stores of all kinds, stacked to the ceiling with products dedicated to them, to prove it. Among these products, food plays a very particular role: wet, dry, in tins or in bags, diet or natural, tailored for a particular size/breed or to clean teeth. The meaning is already there; it imposes itself as proof. The objective isn't to decipher it, but to establish it within a condition that allows for signification (Greimas 1970), finding a metalanguage that allows us to describe the way in which it is produced and, most importantly, how it changes. The growth in knowledge, therefore, shifts: it is not about the sign but about what comes before it, the operation that places it within a culture allowing to change its cultural and social meaning, to negotiate and produce other communication and, later, new meaning.

The animal is not the object of zoosemiotics as discussed here, neither is its language nor any other system of signs that affects it directly as a living creature that communicates, but *animality*. By this we do not wish to refer to a presumed Aristotelian essence, but to that collection of processes that forms the basis for our identification of the animal and its differentiation from the human. Not if and how a dog talks, but why it might be possible to be interested in such a thing, posing the other, the barbarian who produces "unpleasant and unarticulated" sounds (according to the etymology of the word), as someone who is worth getting to know. It is according to an idea of animality if we decide, for example, to eat certain species and not others, keeping some very close to us, perhaps even within the confines of our homes (the domestic animal no less), and others further away, in a pen. Not to mention those who live in the forest around us, and how in order to defend ourselves from them, we build a new pen into which we also place ourselves. Food, distance and the social and familial roles are all products of an idea of the animal and the reason for which we consider other living creatures, inserting them (or not) within our own community. These are forms of cohabitation that end up changing us (and our idea of humanity) as much as the animals (Despret 2002), along with the entirely semiotic concept that is nature (Marrone 2011, 2012).

Anthropology has long shown us (Descola 2005; Viveiros de Castro 2009) that the idea an entirely unique concept of nature exists is a naivety shown by each and every culture, the Western one in particular. By considering themselves necessary, cultures pose an equally necessary idea of nature that soon becomes an ontology, a way of understanding reality that acts as a paradigm within which not only animals find their place, but all elements of the environment that surrounds us. Why is it that, for some, a pig is a filthy animal to be kept as far away as possible and never to be eaten, whilst for others it is not? Are we sure that a dog is more suited to domestication than a wild boar? And why should we look after the planet? Because we see it as a living creature or simply because its resources are dwindling? It is evident that the answers to these questions vary enormously in time and space, demonstrating how nature in general, and animality in particular, can take different forms depending on the culture in which they exist. This will form the nucleus of our investigation: not animality but *forms of animality*, the forms through which the concept of animal, and the dog in particular, is constructed and transformed over time, but, above all, the way in which it is translated from one object to another, from one text to another. Because if it is true that those discussing animals are a broad collection of empirically given entities, what counts as far as we are concerned is the coherence of that collection or rather that discourse that, in no way paradoxically, talks *of the* animal but, above all, talks *the* animal.

2 From Prey to Friend

If there is one place in which the forms of animality are plain to see, it is in those areas that focus on their nutrition. It is a complex issue, not least because animals are forced to take on different roles, the first of which has the animals themselves as food. Before humankind was able to sustain itself with the invention of agriculture and livestock, most animals must have been little more than a lucky meal, limiting the distinction between species to an issue of danger and, of course, taste. Behind that which Ingold (2000) chooses to call "procurement" activity, there is always a way of perceiving the environment guided by knowledge and techniques. The hunter-gatherer communities moved rapidly in small groups, in search of food and shelter, and among them there couldn't have been much room for creatures which could have been considered as a possible meal. According to a hypothesis by the father of ethology Konrad Lorenz (Celli 2001), in his book *Man Meets Dog*, it is at this point that something began to change (Lorenz 1983).

There are no certainties but, Lorenz suggests, this may be what happened. The hunts undertaken by the men must have taken place in the morning, with the light in their favour, as by dusk they would have been eager to find shelter. The night wasn't suited to human sensory apparatus, and the only way to defend oneself was to barricade themselves in. As much of the prey was eaten as possible and the carcass was left behind. Dogs were the ones to benefit from these leftovers, and a pack, which had understood this, learned that by following those strange creatures, they would be able to get a meal without any effort. They kept a short distance from the humans, both in order to follow their tracks and, in cases of necessity, as a potential meal (for the dogs), though they would have been particularly difficult to catch. The dogs' senses were perfectly adapted to darkness, their hearing and sense of smell allowed them to know in advance when dangerous animals were approaching and, when this happened, they would become agitated, barking and whining. It was at that point humans must have understood that those irritating creatures could have a different use as a useful and sophisticated alarm system. In order for the operation to work, however, the pack had to stay close to the group. The ethologist suggests that it was at this point that one of the humans must have had the idea of throwing one of those creatures something to eat. An extra piece of meat, perhaps, caught with exactly that in mind. Suddenly, dogs were no longer just a potential meal. Now they were being fed by humans, who used their gifts to keep close to them the creature that mother nature had blessed with such preciously accurate senses. The dog offered its skills to the human, and they repaid the favour with food. More than a gift, the food was in fact a counter gift, extended to build the bond of loyalty that is today signified by that very animal.

It might be said that Lorenz' theory is nothing more than a story; it could have gone differently, and it could have been children or a particularly sensitive individual that first adopted a puppy in order to play with it. We don't know and we never will. However, what must be saved from each of the possible stories is that the relationship between man and dog (as is the case with all domesticated animals) is about food: who provides it, who can become a meal and under which conditions, and which, like sheep, cows or chickens, beyond being food, also produce it. Also central is the way in which these creatures are fed. There is always someone who holds out a gift and someone else who reciprocates; and food is often, if not always, a counter gift rather than a gift. It arrives as a response to a gesture, potential or real, and it is the nature of this gesture that we wish to investigate.

3 The Nutritional Pact

Between man and dog, there exists that which we could call a *nutritional pact*, an entirely implicit agreement that, through food, binds them. We are not saying that feeding this faithful friend is the only way in which to build a relationship between them and us but that in one way or another, such a relationship gains significance through food. How then can we find out the terms of such a pact? How to rebuild the many logical possibilities that could govern it? How do we understand what a dog is to a person and vice versa? Two ways of facing this question are evidently ethology (Alleva 1994) and anthropology, which attempt to resolve the problem by, on the one hand, looking at the animals and considering the way in which they behave, learn, understand and express themselves (Mainardi 2011) and, on the other hand, by looking at the human communities and the role that these creatures, at times, play within them (Descola 2005; Despret 2002). The semiotic approach, however, though it recognises the importance of such points of view and attempts to deal with them, focuses its attention on one of the areas in which such a relationship is made evident and, therefore, articulated, a social discourse that, for its success, has to pay a great deal of interest in rendering explicit and delving into (albeit in its own way) the forms that this pact can assume, continuously exploring the various possibilities and paying particular attention to understanding each tiny change. Such a social discourse is advertising. Besides being a marketing tool, in fact, advertising should be considered the product of that mass culture that in 1964 Eco was already considering as an object of investigation for human sciences.

Selling dog food means, above and beyond anything else, selling a relationship with one's animal, not least because in this case the buyer and the consumer are not one and the same. Adverts generally articulate the meaning of objects; it confers value upon them by inscribing in them the very meaning that will lead someone to buy them (Floch 1990; Marrone 2007). In our case in particular, such meaning cannot help but coincide with the relationship the buyer maintains with the real consumer: that sweet, eternal puppy to whom advertising means absolutely nothing but about whom adverts talk incessantly. Furthermore, the companies, having to stay competitive whilst coming up with an ever-expanding product range, are continually forced to invent new strategies, exploring the multiple dimensions of that relationship between a dog and his master without which they would not be able to build their own image.

Before we start, I would like to give a brief note on methodology. The corpus of advertising that we have looked at includes more than 150 commercials of lengths varying between the canonical 30 s and 1.5 min in English and Italian for a market that includes Italy, the United States and the United Kingdom. The majority of these were made in the last 10 years, but our desire to investigate the changes experienced by a relatively young market such as that of dog food has led us to include a certain number of adverts from the 1960s and 1970s. We have been guided by semiotic models, in particular those regularly applied to advertising discourse, which provided an initial, necessary criterion of relevance with which to articulate the market. We make particular reference to the so-called *square of consumption values* elaborated by Floch (1990), which we will discuss in the following paragraph. Due to space restraints, we will limit ourselves to present only those adverts that we believe to be the best examples of any given category.

4 From Values to Valorisation

The semiotic square of consumption values is a methodological tool used by Floch (1990) to analyse the semiotic strategies used in advertising to promote any kind of product. The model is achieved by starting from the assumption that advertising is a kind of discourse whose aim is to give meaning to the products themselves according to the social role that they can assume. It is important to emphasise the originality of such a perspective in terms of the common vision based on the concept of value. The values attributed to a car, for example, are clearly different to those attributed to a food product (Mangano 2014). If in the first case one could highlight its functions, comfort, safety, prestige and so on, in the second one would discuss nutritional value and health benefits, alongside the obvious question of flavour and, therefore, enjoyment. The semiotic perspective, which focusses on formal procedures through which the values are differentiated, allowing them to be perceived within precise cultural systems, comes above all of this. When a car commercial (Mangano and Marrone 2015) takes great care to point out how a particular model performs excellently *and* consumes very little, or to highlight the car's prestige *in spite of* its modest price, it is usually because these values are perceived as different, even incompatible. In any case, they are organised according to precise relationships. In order to pinpoint such procedures and such a cultural system, Floch looks at what each commercial is, that is, a story in which the product contributes in a decisive way to resolving the protagonist's problem as raised by the story (a "lack", to borrow the terminology used by narratologists) and which kick-starts the tale. It is indeed narratology that tells us how, within stories, an object can be presented as an endpoint, an objective that when reached leads to the fulfilment of the subject, or as a vehicle, a kind of aid without which the hero can never reach his full potential. If we think about it, the same thing happens in advertising stories. Thinking of a car

Table 1 The semiotic square of consumption values by Floch (1990)

Practical advertising		*Utopian advertising*
Use values		Basic values
Negated basic values		Negated use values
Critical advertising		*Ludo-aesthetic advertising*

as a vehicle – or, to use semiotic terminology, according to a *use value* – means insisting on how comfortable it is, on its safety features, on all of those values that make it the best "means". Considering it an objective – a *basic value* – means, instead, insisting on the way in which it enters into a relationship with the subject, contributing to the definition of his or her personality, for instance, in terms of his or her social status. Something very similar happens with food. A yogurt can be presented as way to improve the activity of a slow intestine, whilst another can become the symbol of a healthy and active lifestyle. Values put forward vary enormously; what does not change are the procedures that allow such a valorisation to happen.

Taking this qualitative opposition as our starting point and considering it as a semantic category (Floch 1990), we will therefore also acknowledge two logical positions that directly descend from two precedents and which the semiotic square allows us to visualise. They are terms known as sub-contrarian, and they come from the negation of the two original positions, that is, from a new type of logical opposition, this time privative. We will therefore have a *negated use value* every time the instrumental value is negated without necessarily affirming its opposite. Finally, from the *negated basic value*, we will have a logical position from which we attempt to highlight those characteristics that will allow us to understand the product as worthwhile according to a form of reasoning that is posed as rational.

Speaking of operations of valorisation and not values changes the perspective from which we look out at the product and the communication. The value is an effect of meaning that realises the way in which the text proposes and articulates the valorisations at play. If the values change over time, the valorisations remain stable, meaning they become an important communicative variable on which a communications campaign can be built. Schematically it is as follows (Table 1).

5 Valorising Dog Food

We can easily see how the habit of giving dogs the leftovers from human meals should have lasted for millennia. When the industry realised that it would be able to exploit the particularly indulgent tastes of these animals to give economic value to

Fig. 1 Commercial for *Gravy Train* (1970), whose human protagonist does all she can to show us the speed with which a dog's meal can be prepared using this product. A good example of critical valorisation

Fig. 2 Commercial for *Purina-Friskies*. Here food is never discussed and the dog is presented as a master of life. This is, without doubt, a utopian valorisation

human leftovers, it was already the latter half of the nineteenth century. It was an enterprising electrician from Cincinnati, USA, James Spratt, who around 1860 set up the first dog food company. His vision took a rather unusual form. Not that of a generic tinned "mush", as would be the case some time later, but a biscuit very similar to that eaten by humans, children in particular. Another thing Spratt understood immediately was the value of communication. A real marketing campaign using cigarette cards was used to launch his *Dog Cakes*. The target audience was rather well-off, and the idea of feeding one's dog with biscuits could only take root in a social class for whom these animals were an indispensable hunting companion. For dogs, food continued to be a counter gift, whilst for human beings, the four-legged friend became more like a child (Mangiapane 2018).

However, when the animal food industry took off, it did so starting from diverse assumptions. As advertising clearly shows, it quickly became clear that dry food and tins would interest the largest number of pet owners, people looking for a quick way to solve the wholly human problem of what to feed the latest addition to the family. In the 1960s and 1970s, particularly in the Unites States, there were a vast number of adverts that regularly showed the moment of preparation. So, whilst excited dogs followed the product or tried to jump onto the table where it was resting, we saw women showing us that you simply needed to add water to transform some miserable, meagre pellets into tender "pieces of meat" (Fig. 1).

This is evidently a case of critical valorisations, in which the continual reference to time and simplicity, alongside the clear demonstration of how to prepare, comes back to this desire to insist on convenience. If the birth of animal food came about as an aristocratic game, its democratisation was driven by efficiency and rationality.

To help us understand the profound transformation in the way we think about dogs, we will now introduce a commercial from the brand Purina-Friskies (Fig. 2).

Fig. 3 Commercial for *Purina Pro Plan*. The world depicted in this case is entirely canine, and the product becomes the fuel needed to allow the dog to do whatever it wants. Here the valorisation is practical

It features three characters: a little girl, her father and their Labrador. The scene takes place in a country lane, where the man is throwing a ball for the dog. The shots make abundant use of slow motion and extreme close-ups as they alternately follow the ball and the conversation that is taking place between father and daughter. We then hear a voiceover saying:

> If he could talk, he would tell you to follow your instinct. // When you see something you wish was yours, go and get it. He would tell you to be brave, to never be afraid of what you desire. // Follow it. Don't give up. Never give up. // And when you finally get what you desire, and your heart is filled with joy, pride and satisfaction, he would tell you to let it all go. // But he cannot tell you anything he hasn't already shown you (translated from the Italian).

At no point in the commercial is any allusion made to food, not verbally nor visually. But this isn't the most striking element. Signalling the spirit of the times is not just the disappearance of the product, but the fact that the father limits himself to simply verbalising what the dog is teaching the child. He is the master, the guide. And he takes on this role not because he doesn't speak ("he can't find the words"), but because he can't speak, because of the good example he gives. If, as we have said, food is always at the centre of the relationship between human and animal, here we have reached a point when it is taken for granted and explicit reference is no longer needed. The human-animal relationship is depicted, in essence, in order to "say" food, giving it back its most profound meaning. No pellets or tins, but the utopian sublimation of it all.

It is Purina that provides us with yet another good example of practical advertising – a sign that nothing forbids the same brand from suggesting a number of valorisations for their own products, often contemporaneously. The campaign in question this time is *Pro Plan*, produced both for the American and the Italian markets. Setting this commercial apart from those we have already seen is the fact that there are no humans present here. Only at the very last moment do we see the legs of a young woman, but only to call the two starring dogs for their walk. Up until that moment, they had been lying on a sofa, sleeping (Fig. 3), whilst the video drew the connection between the involuntary movements of one dog – paws that move, little barks and ears pricking up – with its dream, alternating real-world frames with those from the dream sequence. In its reverie, the dog has to retrieve a ball that has ended up in a lake. After a short run along the boardwalk, he throws himself in without hesitation and begins to swim as the shot moves below the water line to

Fig. 4 Ludo-aesthetic commercial by *Cesar* in which the question of taste is thematised by humanising dog food

show the short legs moving rapidly. It is here that we see the courageous dog is not alone. Next to him is the other animal that we have previously seen on the sofa. The two dogs couldn't be more different: the dreamer is a diminutive, stocky dog with short legs, whilst the adversary is an elegant Irish setter. The setter is ahead when the first dog enters the water but, in this new condition, their positions of strength are inverted, and after a mighty swim, the little one triumphs, holding the coloured sphere. Through all of this, the scene is accompanied by the aria *Largo al factotum* from Rossini's opera, *Il Barbiere di Siviglia*, which, obviously, reaches its zenith when the ball is reached. At the end, when the shot moves from the dream sequence to reality, silence falls and a voiceover explains the nexus between what we have just seen and the product: "If your dog can dream it, Purina Pro Plan can help him achieve it". And then, reiterating the payoff, we see words appear under the producer's logo, "Nutrition that performs".

It is evident that the universe we are faced with here is almost exclusively animal. The little dog dreams of winning the competition against his tall, handsome and elegant stepbrother, whilst the human, and his relationship with the animals, is relegated to the sidelines. The human is the one we assume has thrown the ball, but most importantly, he/she is the one who buys the right food for their faithful friend so that it may realise its canine dreams. Thus we literally enter the animal's mind and realise we need the product, which poses itself as a vehicle, a use value, in order for it to fulfil its desires. But it is the animal that needs it for its own reasons, not us.

In all of this, the reader will have realised, we have missed an aspect that, when it comes to food for humans, usually takes centre stage: taste. Furthermore, the conditions are entirely different. Here it is taken as a given that the human will buy a product that he will never taste and on which he will never receive articulated feedback that would allow for a more developed judgement than the plain I like it/I don't like it and I'll eat it/I won't eat it. And yet, just as with human food, taste is also discussed here. Let's take a look at a new advert.

This time the protagonists are a West Highland White Terrier and its female owner. The brand is Cesar, whose logo bears the snout of the same small dog (Fig. 4). The scene takes place at the table of what seems to be a bistro, under the porticoes of a typically Italian city with plenty of passers-by. The dog seems over-excited, continuously jumping up at his owner's legs in anticipation of his meal. A female voiceover says: "I love that my dog is always so full of life, that's why I give him Cesar every day". There is then a cut and the voiceover begins to say: "Delicious steamed meat, healthy vegetables and cereals", whilst we are shown a wooden chopping board filled with small chunks of steaming hot meat, carrots cut into small cubes and peas (Fig. 4). The ear of wheat at the bottom seems to hint at the best-left-unidentified cereals they are referring to. The image, however, is not one of a tin. There is no gelatine, and most importantly, no indistinguishable mass, as is usually the case. From synthetic cooking, to use Montanari's term (Montanari 2004), we have moved to the analytic, in which every taste is easily identifiable. Or that's at least what they want to be the case, so that the visualisation of the tin's contents is perfectly compatible with a human aesthetic. The instinctive annoyance that we feel when faced with animal food is completely anesthetised, to the point that we could actually consider tasting what we are being shown. But it is only for a moment, because 1 s later the board moves to a perfectly white plate on which the morsels are laid out in a pattern recalling that found in the tin from which they have come. At this point the human appetite comes to a standstill, as the dog launches itself at the plate to devour its contents. The voiceover says: "He loves it, and I know it is good for him". "Cesar, to love him like he loves you". Once again, we have the payoff.

It is important to notice how, in this case, it is not just the food and its visualisation that suggest how delicious the product is, but the entire context in which its consumption is inserted. We are not in a domestic environment, that of a common meal, but in a bistro, in which food is assumed to be extravagant. It almost feels as if the owner has invited her dog, not just to eat out, but to eat something "human", a meal that wouldn't ruin a dinner table, as confirmed by the plate on which the contents of the tin are served. And then there are the images that, at least for a few fleeting moments, speak directly to us, making us visually taste that which the dog would usually be the only one to eat. Here, the utilitarian dimension of food, its role as "fuel", is denied, instead affirming a more hedonistic, ludo-aesthetic dimension, linked not only to enjoyment but to perception also.

6 Utopian Forms

Though the grid of values has given us the opportunity to explore the principle ways in which one can discuss dog food, it is clear that the issue of meaning in this product doesn't end here. Rather, a categorisation such as that offered by Floch must be seen as a starting point, a map from which the exploration begins. Every valorisation throws up questions concerning both generic and specific aspects. For example, once a valorisation strategy has been chosen, how do we transfer that to the screen? We have established the role that the product must play, but we must imagine the story within which this can happen. Why choose a bistro and not a restaurant? Why a woman and not a man? And, who is the woman in question? What does she do? Is she an entrepreneur or a housewife? Is she walking the dog in the park or taking it to a dog show? But above all, how is the food represented? Is it devoured or savoured? Is it eaten next to the owner or not? On the floor or at the table?

As Marrone (2001) explains, profound values are one thing, the narrative programmes through which these values are evoked and characterised are quite another.

If adverts have the aim of building an imaginary world around the product, and if this must be conceived strategically in order to differentiate itself from that of competitors, there are a thousand ways in which to give it depth, to characterise it with those details that make it credible. Recognising advertising as utopian, highlighting macro-aspects that differentiate it from its opposite (practical) or its contradictory (critical), is just the beginning. If, initially, it is enough to say that the basic values should be seen as a final value for the subject, and that the union with the object constitutes a form of realisation, it is crucial to understand exactly what is happening when the connection is made. Is the subject delighted with simply attaining his or her objective? Is there a change in his or her social status? Or is this change something more profound, like the subject's perception of his or herself, his or her own identity? In utopian advertising, it seems that what is at stake is the nexus between *doing* and *being* (Marrone 2001), between plans for action and a subjectivity constructed from passions and sensations, corporality and perception on which the action itself is necessarily based. Generally, it isn't even a given that connection is made between subject and product within the advertising story, often the object remains an object of desire. Just think of car adverts. We contemplate those cars and we are led to see ourselves driving them, but the commercial does not give in to our textual simulacrum, pushing us so that the repression of this lack might take place only in the real world. Then we have all those cases in which the object – a mobile phone, perhaps – is left to one side, forgotten, or simply used to do something else like take photographs or listen to music. Once it has been able to transform the subject, it loses all importance. In the case of dog food, it is impossible for the connection to be made. Buyer (human) and consumer (animal) are not one and the same, and the only way to consider the realisation of the subject-buyer is not through a nutritional route, but through an existential one. For this reason, utopian advertising represents a fundamental communicative tool for dog food, and this is why we find in our corpus a great number of diverse ways in which to articulate this valorisation in a narrative way. As Marrone (2001) says, there are different possible narrative prospects for each of the positions of the grid, some that focus on the subject, others on the object and then also the relationship that binds them. If animal food becomes an area of unlimited experimentation for advertisers, for us, showing some of the results of their creativity is a way to rebuild that very relationship between human and the animal in which we are interested.

7 Human-Dogs

Two of the most interesting examples come once again from Cesar, featuring our West Highland White Terrier as protagonist. The first begins with an unusual scene: an enormous glass-fronted building out of which, in the evening, people in suits emerge. There to greet them cordially is the night porter, who quickly finds himself alone in the enormous building, ready for another night without sleep. Once the doors close, the man moves towards the lifts whilst whistling for someone to join

Fig. 5 The transformations of the subject in a commercial for Cesar

him. We only discover who when the doors are about to close and the little white dog appears, passing in between the man's open legs (Fig. 5). To tell the truth, it is a rather strange choice of guard dog: small, hairy, seemingly inoffensive. Wouldn't it be better to bring along a large Doberman or gigantic Molosser? The reason for this choice is made clear in the following scene when, after having been round to turn all the lights off, the little dog and his master slowly begin to do something unexpected. The dog is the one who starts off, throwing a roll of tape onto the floor so that his master sees it and picks it up to play with it as if it were a ball. By the following scene, the chase has begun. And so, whilst the background music (a waltz in 6/8) becomes increasingly upbeat, the pair start to play with everything: scooting about on the chair on wheels, throwing shredded paper into the air and, of course, playing with the ball, all the way down to the garage. Then, they take a spin on the cleaning cart and another through the metal detector, and so on: everything that you would have ever wanted to do in a closed office but never dared. This continues until dawn, at which point the master pulls out the classic packed lunch. And when the dog barks to claim his own, the tin containing the product is shaken before being turned out onto the usual white plate and taken to the meeting room. Here, with an eye on the skyline outside, the two of them serenely eat their meal, one next to the other, both leaning on the same surface. There is no explanation, no voiceover: the images have said it all. All that is needed is the logo with its payoff: "Love them back".

Here, the subversion of a stereotype has a value that is of great interest to us. The dog is not of any material help to the guard, he plays no role in scaring off possible threats, he is not a weapon – at least not in the conventional sense. The real enemy here is boredom, and the puppy is the only one that can help to defeat it, turning his master into a child. This is the power of our four-legged friend: to modify the human's perception. Transforming a roll of tape into something to be chased, the metal detector into a toy that emits lights and sounds, the cleaning device into a go-kart, and so on. Pure bricolage (Lévi-Strauss 1962; Floch 1995). Boredom becomes an opportunity, repetition becomes discovery, solitude becomes freedom. And when the food arrives to sanction the relationship between dog and master, the food that the pair eats contemporaneously and from the same surface, is a counter gift, but one offered in exchange for a way of life, a gateway into an existential dimension that the dog makes possible with no regard for reality. Furthermore, whereas in many adverts the dog is presented as a kind of child, here the opposite is true: the man is the one who changes. The utopian form is declined to begin from the Subject, who we see transforming before our very eyes, to the point he becomes somewhat dog-like himself when it comes to finally biting into his sandwich.

8 Dog-Humans

At this point it would be interesting to analyse a new commercial for Cesar, which is part of the same campaign and which should, therefore, be in line with such principles, adhered to an identical strategy. We will see how it is enough to change just a few elements in order to make a utopian valorisation (and therefore to a human-dog relationship mediated by food) for it to take on a very different form. This time the setting is that of a small town located on a typically Italian mountainside. A rather elderly man opens a shutter in his pyjamas and, behind him, we immediately see the dog-logo at whom the man smiles. The music (a romantic étude), played on the piano, is particularly melancholy. We see the man descend the stairs and put on a flat cap in front of a mirror. A glance at the dog is enough to make him change his mind and change it for a bowler hat. The next shot sees the two leaving the house and going to a bar, where a woman serves them breakfast: coffee for him and a bowl of water for the dog (Fig. 6). What is interesting is the fact that these are consumed on the same large wooden table, one next to the other. After complicit looks and synchronised head movements, the two leave the bar and begin their walk, which takes them to the florist in the town square. From here they go straight to the cemetery. A new frame shows a brief moment to adjust his tie and the dog's collar with its medallion, and the two move towards a grave to leave flowers. The man's look instantly tells us that his wife is buried there. There are a few moments of sentiment-filled silence, and the dog licks the master's hand, who responds with a smile. This is the signal (given by the animal!), they are to go home to eat. We then see the tin containing the product being pulled from the larder and, after a brief cut, two plates next to one another. In the first are cured meats and cheese and, in the second, the contents of the tin. The two eat outside, with the man sitting on a bench and the dog by his side on a step.

As we can see, there are many of the same elements as in the previous advert, the meal together, on the same level, both on the same sort of plate, and the complicit relationship, even a physical similarity between dog and master. Only that here the utopian conjunction to which the product bears witness occurs from a very different

Fig. 6 The transformations of the dog in a commercial for Cesar

transformation. It is not the man that becomes a child thanks to the dog, who somehow brings him to his level; here the opposite occurs. From the beginning, the dog is something other, quite what is only revealed when they are both at the grave: he is now the old man's companion. Looking at their relationship, we effectively see the relationship that had existed between this tender husband and his wife. This explains the other meal they consumed together, breakfast, from which the product was entirely absent. The idea of the product as a sort of counter gift is watered-down: there is no prize, just the habit of a continual sharing that can be linked to Cesar, or not. This is an involuntary paradox for the advert, which, if we look closely, is making the point that it doesn't matter what you eat; what matters is eating it together.

9 The Subjective Correlative

We are now used to the fact that in adverts for dog food, the relationship between product and buyer is mediated by the figure of the four-legged consumer. Now, in the last cases we have seen, adverts have insisted on thematising, articulating the relationship between the buyer and consumer, insisting on the transformations that interest those in it. So, if in one case the man became a child thanks to the dog, in the other it was the dog who became a human, substituting the companion that the old man no longer had. In both cases, the change takes place on the level of subjectivity, but if we are told about it, it is because it cannot help but affect our relationship with the product. It is no longer a case of feeding something to our faithful friend, but to provide for the needs of a collective actor who has come to determine itself and bring together the human and the animal. As a consequence, the object-product relationship suddenly goes from being mediated to being direct, linear. This is the case at least from the perspective of the signification processes taking place. But it is worth repeating, none of this happens because of the food's quality or taste, but because of the subject that consumes it. Unlike that which happened in the first commercial for Purina Friskies (§ 5), and as is the case with much of our corpus, the choice isn't making the food disappear, thus allowing its supposed importance to hail from the relationship between man and dog. Here, the food is seen and eaten: it is who eats it that is different. The tin is no longer the *objective correlative* of affection towards the animal, to use Eliot's famous expression, but the *subjective correlative* of their shared feeling. We are, therefore, a long way from the "humanised" vision of food proposed by the ludo-aesthetic advert, trying to make it appetising to us. Here they work on the consumer, but they do so *through* the food, exploring that important aspect which is *conviviality*. Provided that this concept is understood not as the simple sharing of a meal, but as something more profound, which has a political, social and identity value (Illich 1973). Food, people and everything else that contributes to the characterisation of a shared experience are transformed in the name of an action that starts from the subject's sensory channels. It is as if, suddenly, we could go back to that common sensory field, to borrow a phrase from phenomenology (Merleau-Ponty 1945), in which everything is possible, even that a dog and a human share the same subjectivity.

10 The Society of Dogs

Up until this point, the relationships' *mise en scène* by advertising have more or less directly involved a master, an animal and the food that – present or not – they must both eat. However, things can be much more complex, from the moment that each of the characters involved assumes further relationships with other elements. The case of Pedigree Pal proves this very point (Fig. 7).

The advert, shot almost exclusively in black and white, begins with a lock clicking shut. As the shot pans out, we see that it belongs to the reinforced metal door of a typically American style colonial house. It is a normal house that could be in any part of any city. Closing the door is the owner of a large, short-haired dog that he is now walking along the street, held by a lead. There are no romantic Italian towns or nice bistros, no forests or lakes, just an oppressive city landscape, with people intent on getting on with their own business. The man is irritated and he gives a dirty look to the car that won't let him out, pulling the animal to a halt. At one point the man, white, late middle age, sees a young African-American man at the other end of the street, also walking his dog. In a shot and counter shot, we see the two look each other up and down, with a look that evokes memories of that racial tension that has never been fully resolved, particularly in the American inner cities. When the two meet, the hostility is palpable, but not for the two dogs who, having nuzzled one another, start licking each other's faces. It only takes a moment, the men's eyes lower to look at the animals and when their gazes return, they are different. The white man smiles and so does the African-American man, who also starts talking: "It's a big dog". "Yeah", the other man answers, "his name is Patton". "This is Murphy". "Murphy! Nice to meet you, Murph!", says Patton's master, stroking the animal's head. Then, continuing his sentence, he exclaims, "Uh, nice meeting you!" At that point, the African-American man holds out his hand towards the man (just as the other had done first with his dog) and says "You too". It takes the voiceover to bring us back to the product: "Dogs bring out the good in us. Pedigree brings out the good in them. Feed the good". All of this happens as we follow the African-American man (and Murphy) home, where there is an enormous sack of dry food

Fig. 7 Social transformation in a utopian commercial for Pedigree Pal

waiting. An enormous cupful is pulled out and poured into the bowl before being greedily devoured by the famished dog. We can already see how this commercial seems to differ from the others. Paradoxically, it is unusual to see the dog food in its everyday form, even more so to see it in a perfectly normal bowl on the floor. There are no inversions or transformations; here the dog is a dog. For the white man, about whom we know more, the dog is obviously a weapon (he is called Patton after the general), whilst for the African-American man, the dog is a friend to feed, but the food stays on the floor.

What counts here are the relationships, rendered complex by the fact that there are of three different kinds: the inter-subjective ones, between human beings, which start out conflictual and then become friendly; those between the animals, which are positive from the outset, and very human, in the spirit of conviviality; and last of all those between men and animals which, though they are in some ways the ones that are less thematised – up until a certain point we only see two people walking their dogs – in others are the most important. As the slogan clarifies, it is dogs that bring out the good in us. They are the ones that transform the subjects, but always within an intersubjective framework in which it is the relationship between the dogs that transforms the relationship between people, rather than a single dog changing his master. We learn, therefore, that the *society of dogs* is better than that inhabited by human beings, provided, of course, that the beasts are fed with the product that publicises how it "brings out the best in them". A sort of inverted *animal farm*, in which the doubt remains as to whether it is the product that ensures the story, doesn't end like the one written by Orwell. On the other hand, "Feed the good" means also "give them something good to eat", a transcendent instance that, in fact, would place the dog as the semiotic addresser of an ideal human community. If, as Latour (1999) hypothesises, the politics of nature are linked to the constitution of a collective, it is not interesting only to see who can actually take part (why a dog and not a mosquito?), but also in what capacity. Placing the dog as the addresser means attributing to that dog the sum of responsibility for the value system on which the community itself is based.

11 Conclusion

Such an articulated theme and vast corpus, like that which we have collected, requires far more time to be explored thoroughly. Our aim was to offer the reader an idea of the semiotic approach to the issue of the human-animal relationship, demonstrating how an apparently simple and innocent text, such as an advert, can reveal aspects that are anything but straightforward. It seems to us that there has been a notable emergence of a tendency within mass culture to problematise the role of the animal, and the dog in particular. What is striking is the variety of ways of articulating such social roles, which change dramatically within the same company and even within the same campaign. From this we gain a comprehensive picture that continually redraws not only humanity and animality but also the worlds to which these creatures, in principle, belong. If we think of the natural world, which should be the dog's domain, and of culture, created as it should be by human beings, the texts demonstrate that things are far more complicated than that. This hypothetical separation is continually denied in the name of a transformation that affects both humans and animals, involving the whole of society. Advertising seems to resist the dominant naturalist positivism that considers men and animals as physically similar and spiritually diverse, in favour of an *animist ontology*, according to which there is instead a continuity in the interiority and a discontinuity in the aspect (Descola 2005). When considered in depth, it would seem that what works is treating the human-animal relationship as something it actually is: a relationship of malleable significance that, as such, has to do with a large number of artefacts. Of these, food has a central role. Therefore, if the Lévi-Strauss "good to think" (Lévi-Strauss 1962) seems to have pinpointed what differentiates us from animals, we realise by looking at society through the prism of advertising discourse that this humanity we search so hard for does not simply concern human beings. Paradoxically, following the need to continually exert their own creativity, the advertisers end up, knowingly or otherwise, exploring profound aspects of culture. If something is tasty just because someone thinks it before they have even experienced a sensation of pleasure or disgust within their own body, the frontiers of this thought necessarily include other creatures. If, on the one hand, they can become food, on the other hand, we can want to feed them, creating a real relationship with them that begins with food, proving we're not alone in thinking about how good something is.

References

Alleva, E. (1994). *Consigli a un giovane etologo*. Roma: Muzzio.

Celli, G. (2001). *Konrad Lorenz. L'etologo e i suoi fantasmi*. Milano: Bruno Mondadori.

Descola, P. (2005). *Par-delà. Nature et culture*. Paris: Gallimard.

Despret, V. (2002). *Quand le loup habitera avec l'agneau*. Paris: Les empêcheurs de penser en rond.

Floch, J. M. (1990). *Sémiotique, marketing et communication*. Paris: PUF.

Floch, J. M. (1995). *Identités visuelles*. Paris: PUF.

Greimas, A. J. (1970). *Du sens*. Paris: Seuil.

Illich, I. (1973). *Tools for conviviality*. New York: Harper & Row.

Ingold, T. (2000). *The perception of the environment: Essays on livelihood, dwelling and skill*. London: Routledge.

Latour, B. (1999). *Politiques de la Nature*. Paris: La Découverte.

Lévi-Strauss, C. (1962). *La pensée sauvage*. Paris: Plon.

Lorenz, K. (1983). *So kam der Mensch auf den Hund*. München: Deutscher Taschenbuch Verlag GmbH & Ko.

Mainardi, D. (2011). *Il cane secondo me*. Milano: Cairo.

Mangano, D. (2014). L'anima del commercio alimentare. Pubblicità, brand, comunicazione. In G. Marrone (Ed.), (pp 66-100).

Mangano, D., & Marrone, G. (2015). Brand language. Methods and models of semiotic analysis. In G. Rossolatos (Ed.), *Handbook of brand semiotics*. Kassel: Kassel University Press.

Mangiapane, F. (2018). *Cuccioli*. Milano: Mimesis

Marrone, G. (2001). *Corpi sociali*. Torino: Einaudi.
Marrone, G. (2007). *Il discorso di marca*. Roma: Laterza.
Marrone, G. (2010). *L'invenzione del testo*. Roma: Laterza.
Marrone, G. (2011). *Addio alla Natura*. Torino: Einaudi.
Marrone, G. (Ed.). (2012). *Semiotica della natura (natura della semiotica)*. Milano: Mimesis.
Merleau-Ponty, M. (1945). *Phénoménologie de la perception*. Paris: Gallimard.
Montanari, M. (2004). *Il cibo come cultura*. Roma: Laterza.
Viveiros de Castro, E. (2009). *Métaphysiques cannibales*. Paris: PUF.

Pet Food Communication: Notes on the Crisis of Naturalism

Roberta Bartoletti and Giulia Cecchelin

Abstract Since the beginning of the second half of the twentieth century, an important change has been taking place in western society regarding the relationship between humans and animals – i.e. pets, wild animals and livestock. We wonder if changes in progress can be interpreted as a crisis in naturalism, and we will try to reflect on whether, and how, the relationship of naturalism with other ontologies, first of all animism, can be used as a lens in order to understand the sense and direction of the ongoing processes. We will focus on domestic animals and pet food, because food represents a mediator of extraordinary relevance in human-animal relations. We analysed the packaging of a collection of pet food products which are currently on the market in Italy and Europe. The body of material chosen was examined in order to find answers to the following research questions: What pet food values are being promoted by the producer? What kind of consumer, either human or animal, is being profiled? And what kind of human/animal relationship is being recognised and promoted by the pet food producer? The human-animal relation which is prevailingly recognised and appreciated by pet food producers is one of affectionate and responsible control, while the dialogic relationship, which still appears, is both as radically innovative as it is minor.

1 Introduction: Similitude and Difference in Human-Animal Relations

Since the beginning of the second half of the twentieth century, an important change has been taking place in western society regarding the relationship between humans and animals – i.e. pets, wild animals and livestock. We believe that the sense of this

This chapter is the result of the combined research by the authors who compared and integrated their different methodological, sociological and semiotic skill sets; the article was written by Giulia Cecchelin (paragraphs 2.1 and 2.2) and Roberta Bartoletti (paragraphs 1 and 3).

R. Bartoletti (✉)
University of Urbino "Carlo Bo", Urbino, Italy
e-mail: roberta.bartoletti@uniurb.it

G. Cecchelin
Independent Researcher, Bologna, Italy

G. Marrone, D. Mangano (eds.), *Semiotics of Animals in Culture*,
Biosemiotics 17, https://doi.org/10.1007/978-3-319-72992-3_6

change must be placed in a framework which is inscribed in the wider frame of relations between society and nature (Macnaghten and Urry 1995; Latour 1999; Franklin 2002; Bartoletti and Cecchelin 2016). The dominant anthropocentric perspective has been challenged by alternative points of view – in particular by animal rights movements (Tester 1992). The borders which distinguish humans from animals seem to be becoming ever more ambiguous (Ingold 1988; Franklin 1999), and western contemporary society's relationship with animals is, as a consequence, multiform (Marrone 2011) and full of contradictions (Arluke and Sanders 1996).

The actual complexity of this relationship has its roots in numerous conceptions of nature which were formed between the eighteenth and nineteenth centuries (Williams 1980) and in the dualism of discourses orientated towards similarities and differences (Tester 1992). The dominant discourse, of enlightened and rationalist matrices, states that the difference towards the natural and animal world is based on the uniqueness and the superiority of humans, who are unique in their ability to dominate the world because of science and abstract thought.

The discourse in favour of similarity is founded on romantic sensibilities towards the natural world; although it took a marginal position in the nineteenth century, it fuels both the animalist critique of the late twentieth century as well as the progressive affirmation of a new way of consuming the natural world and animals, through sport, tourism and free time activities (Franklin 1999).

In Descola's classification (Descola 2005, 2010), naturalism is the specific non-universal way in which the western world understands and relates to what it calls natural, articulating a particular combination of differences and similarities. As a matter of fact, naturalism should be characterised by a continuity in human and non-human physicalities (they are made up of the same molecular substances and subject to the same physical laws) and by a discontinuity in interiorities, seeing that "for us, what differentiates humans from non-humans is a reflective consciousness, subjectivity, an ability to signify, and mastery over symbols and the language by means of which we express those faculties" (2013, 173), consistent with what we have already pointed out.

In this way the typically western naturalism would seem to oppose an animism which unites different peoples from North and South America and all the way to Asia and Oceania who, despite variations in their morphologies and behaviours, recognise similar interiorities in all living things (humans, animals or plants) and their ability to conceptualise and feel emotions modelled on those of humans.[1] We wonder if changes in progress can be interpreted as a crisis in naturalism, and we will try to reflect on whether, and how, the relationship of naturalism with other ontologies, first of all animism, can be used as a lens in order to understand the sense and direction of the ongoing processes.

[1] The further two ontologies identified by Descola (2005) are totemism, "a mode of identification founded on an interspecies continuity of both physicalities and interiorities" (p. 160), and analogism, "the grouping within every existing entity of a plurality of aspects the right coordination of which is believed to be necessary for the stabilization of that entity's individual identity" (p. 212).

We will analyse the current situation, and the present changes, in the relationships between humans and animals from a particular standpoint: our focus will be on domestic animals and on a field which is particularly relevant to consumerism, food. Within the scope of human-animal interactions, pets are usually placed at the highest level of humanization and domestication. They become part of the home, a place from which other animals are usually excluded, are often given their own names and are culturally excluded from the sphere of the edible. Over time they have become human companions (Franklin 1999; DeMello 2012), and less emphasis has been placed on the pet's qualities as positional goods and entertainment or just as a decoration. More often than not they are treated patronisingly, like children, or as equals. The relationship between humans and pets can be similar to a relationship of subordination, even though it is always an *affectionate domination* (Arluke and Sanders 1996). Food represents, as we shall see, a mediator of extraordinary relevance in human-animal relations.

2 Human-Animal Relations in Pet Food Packaging

2.1 Objectives and Research Methods

We will now take a look at the packaging of a collection of pet food products which are currently on the market in Italy and Europe. They have been chosen because we consider them to be representative of the mindsets of the producers and, indirectly, of social discourses and imagery and, eventually, more extensive cultural changes.

This analysis was carried out by integrating sociological methods of the qualitative analysis of content with semiotic methods of narrative analysis. At the same time, we tried to maintain a connection between the strategies and the narrative techniques of the individual tests which were analysed and the social discourses and dominant cultural categories in which the relationship between humans and animals in advanced modern societies belongs.

The field of investigation is therefore quite large because the pet food market has been in constant growth over the last few years (globally +4% both in 2015 and the year before[2]), which has been accompanied by a relative growth in the number of brands on offer and also an increase in the range of products available. The creation of this corpus inevitably entailed some delimitations: a sample of the European market was taken,[3] and we decided to concentrate solely on products geared towards

[2] See petfoodindustry.com/blogs/7-adventures-in-pet-food/post/5609-global-pet-food-trends-sales-and-volume-rose-4-in-2015; last accessed 25th June 2017.

[3] With very few exceptions, the producers use the same packaging design all over Europe, the only difference being the language. This research could be further extended to reveal if the brands sold globally maintain the same expressive characteristics in other geographical areas as they do in Europe.

cats and dogs, which represent the largest segment of the European and Italian markets.[4]

We have included three existing macro categories of pet food products on the market in the corpus:

- Dry food, sold in various sizes in bags, but which contains enough to cover numerous meals
- Moist food in single serving portions
- Treats, typically dog biscuits in bags or single serving food sticks[5]

The material was collected by visiting different points of sale in both specific pet shops and animal superstores and in more general points of sale (supermarkets and shopping malls).[6] We gathered photographic documentation and annotations of the most popular brands and studied the range of products on offer. We then proceeded to increase the corpus by including examples of products which were not actually on the Italian market, but found on e-commerce sites specialised in pet food.[7]

We then created a sample body inside these restrictions (Greco 2011): by comparing the communicative strategies used on the packaging, we were able to formulate some interpretative hypothesis which was then generalisable to the objects taken into consideration and organised into homogeneous subsets. For each of these subsets, we then chose one or more cases on which we could carry out a more in-depth analysis with the aim of describing in detail the contents communicated from discursive configurations which are typical of that set. The analysis was carried out by using a semiotic method, employing tools which will remain more or less implicit in the argument proposed here, except when it is retained necessary to offer the reader a brief explanation.

The body of material chosen was examined in order to find answers to the following research questions: What pet food values are being promoted by the producer? What kind of consumer, either human or animal, is being profiled? What kind of human/animal relationship is being recognised and promoted by the pet food producer?

[4]According to FEDIAF estimates, 21% of European families owned at least a dog and 24% a cat in 2014, while in the EU the percentages are 18% and 26%, respectively; in Italy the percentages are 22% and 19%. Cfr. http://www.fediaf.org/facts-figures/

[5]Treats have been analysed but have not been included in the research report for summarisation reasons.

[6]Observation period: May and June 2016. Supermarkets and hypermarkets (all in, or around, the Bologna area): *Esselunga Santa Viola*, *Coop Barca*, *Coop Lame*, *Conad viale Silvani*, *Ipermercato Carrefour* (Casalecchio di Reno), *NaturaSì* via Montefiorino. Pet shops; *Ipergiardineria* (Casalecchio di Reno), *Pets* via Belvedere, Santandrea 1930, *Pianeta Animali* via Zanardi.

[7]www.zooplus.com and http://www.petsmart.com, last accessed 25th June 2017.

Table 1 Map of pet food packaging

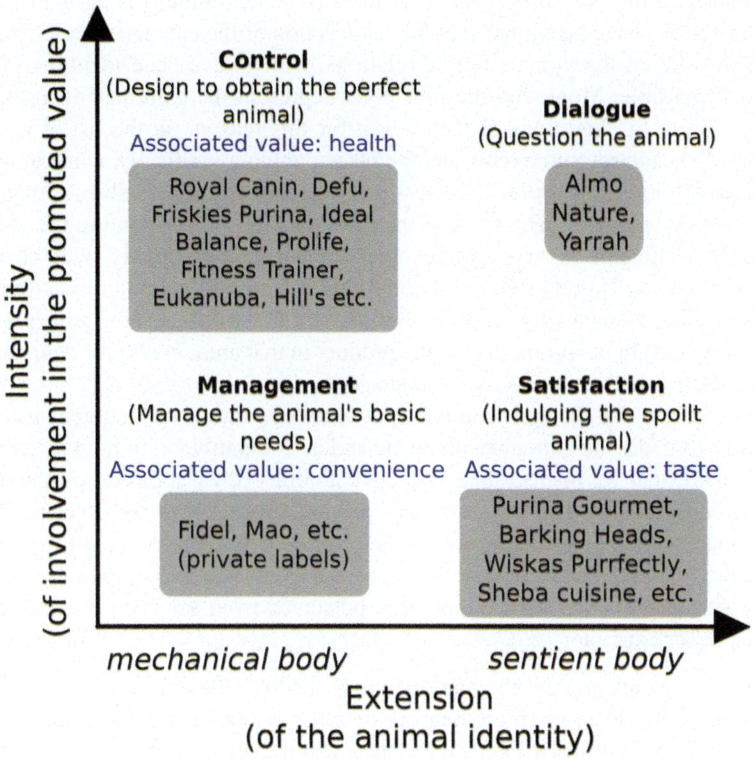

2.2 Relations and Values Promoted in Pet Food Packaging

Predictably, pet food packaging sticks to a narrative structure in which the object of value is not the product itself, but a certain relationship between humans and animals which changes depending on the placement of the brand and the product. In this sense, in a narrative programme (Greimas 1979: ad vocem) which sees the owner of the animal as the subject, the narrative function absolved by the actor-food is to confer a can-do approach to the subject which allows them to shape a specific relationship with their animal. Therefore, more than revealing the values associated with a certain product (which can be, e.g. *health*, *taste*, *convenience*, etc.), we are interested in mapping out the corpus by framing the packaging based on the type of human-animal relationship they are promoting. For this reason the object of the evaluation will not be the product itself (paté, kibble or treats), but the identity of the animal (Table 1).

In this diagram, which is a "model tensif" (Basso Fossali 2008a; Fontanille and Zilberberg 1998) and works as a Cartesian coordinate system, the degree of animals' identitarian autonomy towards humans is measured on the x-axis (the extension of animals' identity), while we see the intensity of the human involvement in the promoted value on the y-axis, or better yet, if the connection between the consumer and the promoted value (the identity of the animal) is strong or weak. Lexicalisations have been placed at the intersection of the categories describing the characteristics of the human-animal relations, which have been identified as precisely as possible. Moreover, the four poles represent the extremes of a semantic field made up of alterations of the values; for this reason, on the list of products attributed to each set, some represent the placement more radically, while others are less intense, or less widespread, expressions of alterations to the values compared to the pole they belong to. The sizes of the four quadrants correspond to the volumes of products which insist on one value rather than on another. Finally, with the exception of the dialogical relation (in the top right), it is possible to identify a dominant pet food value associated with each relation, in other words a value which is more frequently used in the promotion of the product in that area: *health* for *control*, *taste* for satisfaction and *convenience* for management.

It is apparent that the most commonly adopted communication strategy is the one that aims to make the animal as domestic and as compatible with humans as possible. It is not that we neglect their reflective abilities (their ability to do things and think independently when compared to man), but this is not evoked. The producers instead stress the power conferred by the food nourishment to allow the owner to shape the body of his/her pet. In these packs we find the widespread use of communicative expressions which have been borrowed from science (chemistry, biology, etc.), for example:

- Graphics which analyse the ingredients (Hill's Ideal Balance, Prolife, Wildfield)
- Scientific names of the ingredients or of their properties, for example, carnitine, omega 3 and biotin (Hill's Ideal Balance, Prolife, Natural Trainer)
- Expressions like "formula" or "formulated" (Farmina Vet Life) and "system" (N&D, Nutrition System for Carnivores)

The language, the pseudoscientific figurative setting and the insistence on the specificity of the product evoke an intense desire to achieve a specific relationship with the animal, which is moderated by the technical seriousness of the product. Such a relationship is defined as one of control, because the owner uses the food as a way to have an animal which measures up to their expectations, completely under their control. An example of this communicative strategy is Royal Canin's line of dry cat food (Fig. 1).

Royal Canin's product range is quite large and is part of a prevailing trend which aims at the growth and hyperspecialisation of products based on the age, activity and health of the cat (indoor, outdoor, ageing, mother and baby care, regular fit, sterilised, etc.). This line's packaging displays all the same characteristics: a white, uniform background, the logo and name of the brand at the top, the name of the product just underneath, a cat design on the left, photos of some cat biscuits on the right with a pictogram just above which briefly shows the specific conditions of the animal related to the product accompanied by a short description. The cat's paw is outreached, the typical pose of a lazy animal which spends all day inside.

Fig. 1 Royal Canin dry cat food

The rendering is realistic but is outlined in pencil, bringing to mind the dimensioning lines of a technical design, and is accompanied by an unintelligible signature. The same graphic device is also used in the portrayal of the three pieces of kibble in the lower right. Dimensioning is associated with a technical design project, typical of engineers and architects. In this sense, its presence here represents the precision that was taken in designing this particular food. However, at the same time, the "technician" who uses these dimensioning lines is also the owner and can now unerringly control the morphology and behaviour of the cat through food, designing its body until they get the precise measurements they want.

The pictogram shows a cat using its cat box and is accompanied by the text: "stool odour reduction" and "moderate calories". The type of relationship evoked between humans and animals leaves no space for the identity of the latter: it is just another cat and can be represented in a pictogram, thereby eliminating all of its identifying characteristics. The product is also necessary in order to reduce the animal's potentially annoying characteristics, in this case the smell of its faeces, and keep it fit with a reduced caloric portion, thereby making it an element which is compatible with the domestic environment.

The German brand, Defu, is a decidedly atypical example of the polarity of control (Fig. 2). Its entire range of products displays a photograph of some animals (cats or dogs) gathered happily around a person. The human figure, in turn, seems to be looking into the reader's eyes, including them in the scene being depicted. The

Fig. 2 Das Hundefutter vom Bio-Bauern Defu

people here are evidently an iconisation of organic farmers (the Bio-Bauern) who produce the basic ingredients of the product being sold. The suggested script has a bucolic setting: the products from the earth are still collected by hand in the muddy fields and gathered in wicker baskets.

We can see how the cats and dogs belong to the same enunciation, and to the same scenario, of the human figure, from the moment that they turn to look at you; the rhetorical strategy employed therefore produces a curious effect, placing the hungry animal in the same location in which the food is produced: they live in the countryside, instead of in an apartment or in the garden of a semi-detached house, and their owners are the same farmers. There is, therefore, no process of sophistication to be inserted between the raw materials and the finished product. The laconicism of the wording "Hundefutter" ("dog food" in German) is isotopic with respect to the lack of an industrial contribution in the transformation of the product which the packaging is aiming to mean. This is why the *health* value is typically associated with the polarity of control, and here it is implicitly summoned, as a consequence of the genuineness of the product.

We therefore expect to see a specific relationship with the animal, which ties its presence in the life of humans to a scenario which is well-known but outdated: an ancient relationship connected to work, in which dogs were needed to protect the herd and cats were used to hunt rats in the granaries. We still find ourselves

Fig. 3 Fidel, private label dog food

completely inside a relationship of control, underlined by the non-reciprocity of the farmer's gaze towards his animals, which have been given a role rather than an identity.

Instead of being the way to obtain an animal which is suitable for the domestic life of its owner, the food here has the role of evoking a complete scenario: the dogs and cats who eat Defu have no need for collars or leads and live alongside their owners, who gather the food that they cook for them by hand. In this sense the food is attributed the power to fictitiously reconstruct not only the role of the dog or cat but also that of the owner (the reinforced effect of the *embrayage* (Greimas 1979: ad vocem).

The cheaper products and private labels can be found in the bottom left. These products mainly promote practicality through the value-bridge of plenitude. The intensity of involvement is low because the product is valued only as being a cheap, undemanding solution to satisfy a need: the human's only role here is to guarantee the survival of the animal without having to create any kind of relationship with it. The animal at which the product is aimed is not portrayed as having an autonomous identity with which we can measure ourselves against, rather it is stripped of all its individual characteristics, until it is only a name, just like the case of Esselunga's private label canned dog food, Fidel (Fig. 3).

The can is free of images and even the text is quite skimpy. The product has no name, and on the front of the can, we can only see: the logo of the private label

(Fidel), a general description of the product (dog food) and a basic description (with beef and vegetables). On the back we find a brief *consumer guide*, and below this there are some instructions about the quantity to give your dog, based on its weight, and finally there is a list of ingredients. Its financial convenience is never stated, as it is evidently a value which is ineligible in relation to the diet of the pet. It is also worth noting how the modest nature of the pack conveys, relying on an encyclopaedic expertise established for large scale distribution clients, the cheapest possible price. Here the comparison is with DIY home made foods, which entails: the purchase of the ingredients, weighing out the right quantity, their preparation and cooking. Obviously this scenario is never evoked, because this would therefore imply comparing the type of owner who dedicates a lot of time to looking after his/her pet with those who buy ready-made food.

In both the above-cited cases, we have products which are proposed as food nourishment or, even, food fuel. Whether the food is hyperspecific and professional, or generic and simple, it is given to the animal to obtain the standard identifying characteristics: being active, strong and long living. The animal body is, in this sense, treated like a machine, for which the values which have been chosen are assigned by humans: the animal seems to resemble an automaton, able to self-regulate and self-perform, but in need of preprogramming (Basso Fossali 2008a). On the contrary, the following two examples will demonstrate how the packaging represents the *sentient animal* bodies, endowed with a minimum of sensitivity, and even of consciousness.

In the lower right of the diagram, we observe the spoiled animals. The identity of the animal is fully outlined (high extension), as far as it is a type of stylized identitary caricature. The intensity of the involvement is actually quite low, because the animal's identity is not taken too seriously, it possesses a funny, tenderly capricious otherness, and the owner's role in this relationship is only to please the pet. As an example, let us take a look at Purina's Gourmet Perle, a single serving of moist cat food (Fig. 4).

The front of the pack has a photo of a white, long-haired cat, which has been gently placed on an abstract surface, not contextualised in a home setting. Immediately above it we can find the Purina logos, the Gourmet brand and th*e Perle* line. To the right of these there is an image of the product, which is being "served" on a porcelain dish. Underneath the dish we find the name of the product ("Ocean Delicacies") and the image of a shrimp, below this we find the four recipes contained in the pack. On the back of the pack, we again find the image (now enlarged) of the product being served, accompanied by a long descriptive paragraph.[8]

[8] "Your cat is a real expert, and it might be sometimes difficult to tempt it. This is why Gourmet Perle has come up with Piaceri del Mare, delicate recipes which will whet your pet's appetite everyday with an irresistible taste experience. Sophisticated mouthfuls of thinly sliced fish which has been delicately cooked in its own tasty sauce to maintain the flavour of the sea and elegantly combined with inviting shrimp to delight your cat's refined taste buds. Piaceri del Mare by Gourmet Perle, elegant temptations for a unique taste experience, every day. All varieties of Gourmet Perle are available in 85 g pouches in order to conserve the flavour of each recipe".

Fig. 4 Purina's Gourmet Perle moist cat food

The cat in the picture is a white Persian, a particularly precious race: it is not a street cat. It is portrayed as happily gratified: lying on its side, it is gently extending its paw. This is the cat who has everything, which does not necessarily mean it does not want for anything: its stare challenges the beholder to satisfy him. The perfect product for this spoilt, strong-willed cat is sitting right alongside him: served on a modernly designed dish, which conveys the scene of an exclusive restaurant (an isotopy confirmed by the three stars located on the gourmet logo and obviously in the name of the brand), with lots of julienne courgettes. The descriptive text confirms how the procedure being evoked is not just that of feeding an animal but of keeping it in good health. The cat in question is not even hungry, so that "tempting him might be extremely difficult", and eating is not a question of sustenance but is instead an *irresistible taste experience*. The text makes excessive use of adjectives and adverbs referring to a scenario of food tasting which is typically human (delicate; irresistible; refined; delicately tasty; elegantly inviting; elegant, unique), and it does so emphatically that it almost becomes a parody of itself. In this sense, the identitary trait of a *hard to please food expert* is not communicated seriously, as it would be with a human gourmet: the indulgent, amused look on this spoilt animal will be satisfied by its owner.

Barking Heads and Meowing Heads are English pet food brands (Fig. 5); they do not have a particularly wide range of products, but they do specialise in specific products (e.g. for old dogs, dogs with thinning coats or who suffer from a loss of

Fig. 5 Barking Heads "fat dog slim"

appetite). The choice of name and the design of the packaging create, on the whole, an interesting example inside the pole of satisfaction.

All in all the packaging makes an extended appeal to rhetorical figurality, especially when it comes to the name of the brand and its products. The names *Barking* and *Meowing Heads* evidently refer to Talking Heads, the celebrated band led by David Byrne from 1975 to 1991. The choice of font for the logo (like a lumpy old stick) confirms the name's assonance and ties itself closely to the traditions which are typically expressive of punk rock and new wave from that period.

Each product in the range is then associated with a specific name which is itself based on wordplay and tropes: *fat dog slim*, for overweight dogs, evokes the name of the famous musician and producer, Fatboy Slim; *fusspot*, is a typical slang expression for someone who is hard to please (thereby focusing attention on the "taste" value characterised by the pole); *smitten kitten*! or *cat in love*, uses an homeoteleuton to reinforce the diminutive.

Overall, the interpretative work required of the Model Reader (Eco 1979) of the packaging is more demanding compared to what happens for those included in the satisfaction pole: the rhetorical gimmicks being used require an encyclopaedic, erudite knowledge if they are to be appreciated fully. Their disambiguation produces an ironic effect which is caused by the unusual combination of worlds being utilised

(Barking-Meowing-Talking Heads) or by awarding typically human qualities/titles (fusspot) to animals.

We are therefore observing the unfinished anthropomorphisation of the animal to whom some typically human characteristics are being humorously attributed (e.g. to be "smitten"). This semantic scheme is also confirmed in how the animal has been highly stylised, with an almost comic book look, on the packaging. So, although they are geared towards a target market which possesses more highbrow cultural tools, the Barking and Meowing Heads products foreshadow a relationship of satisfaction which is still similar to the previous example.

Finally we have the two unusual cases of Yarrah and Almo Nature, which we have been placed in the top right quadrant of the diagram because both, even though they have a completely different *modus operandi*, share an autonomous identity of the animal which needs to be fed, thereby evoking a kind of symmetrical relationship with humans which is communicative rather than administrative. Their positioning on the right reflects their recognition of the autonomous identity of the animal, while their positioning at the top signifies the high intensity of involvement on the part of the human consumer of the autonomous identity which has been bestowed upon the animal. Compared to the humorous and ridiculous otherness of the strong-willed cat, the animals designed on the packaging here are individuals with whom we have to make an effort if we want to have a decisive relationship, fundamental for the identity of both humans and animals. The two companies also have similar products in common: the price ranges are high. Yarrah's products are organic, while Almo Nature mainly uses human food chain (HFC) ingredients, in other words food which was originally identified for human consumption; therefore neither recognises the classic distinction between human food and animal food which characterises the narratives of the producers.

Yarrah (Fig. 6) proposes images of the same animal on all its dog food packaging: Dano, the Bernese mountain dog owned by Jan-Jaap Roelevink, founder of the brand. As you can read on the back of the pack and on their site, "My Bernese mountain dog, Dano, has been eating Yarrah's products all his life". Like in no other example we have seen, the visuals make use of the face of the company: Dano, with his chest out, seems like he is posing as he stares into the camera. Aware of his role as a spokesperson (spokesdog), he carries his responsibility proudly to guarantee the quality of the product for his peers. He is not just any old dog but a fully fledged identity, unique, just as no dog for its owner is just another dog, and he himself has chosen to eat Yarrah kibble.

Almo Nature (Fig. 7), on the other hand, has chosen a radically different visual: the animals are always in a group and designed in an innovative way. The example shown in the picture is of a packet of dry cat food: the clowder of cats pass from a background in which they are only silhouettes on a white background, to the foreground where they are wearing cat masks.

The mask is a complex device, capable of being both a figure of deception and one of recognition (De Marinis 1982; Elam 1980), but which generally introduces a Thirdness (Peirce 1866–1913), a pathway to something else, inside a communicative relationship. This Thirdness can be achieved in the mimesis of an otherness

Fig. 6 Yarrah dog food

Fig. 7 Almo Nature dry cat food

(disguise yourself as) or in the choice of an identifying trait which is yet to be revealed (just as the ancient theatrical masks showed an expressive quality of the human face which is fully expressed figuratively). In this case the cats wear cat masks, just as humans use masks of human faces to play with the figurativity of those who wear masks for ritualistic or artistic purposes.

These cats have no need to wear human masks to show their own identitary autonomy: the Thirdness that they are invoking is still that of being a cat. In other words, the stratification of faces (one's own face and the masked face) communicates a profound identity which does not need to pass through humanity in order to be recognised, but which is one of complete "catness".

These animals are therefore others which need to be investigated, not just dumb subjects to which qualities need to be assigned. The identity of the animal does not undergo any process of subjection or marginalisation: Almo Nature presents us with a radical identitary distance which it does not try to reduce, but instead sees mystery as a value.

In this area, which has been defined by a dialogical relation between humans and animals, it is impossible to identify a value which is exclusively associated with the product, for example, in the case of *taste* when compared to *satisfaction*. This is not only because we find only two in this set, something which makes it impossible to establish a hierarchy, but also because, in both cases, the packaging presents numerous predicated values compared to the product: *taste*, *health*, *convenience* but also tradition, in the case of Yarrah, and, explicitly and surprisingly, *love*, in the case of Almo Nature ("Pet food + Almore" is the payoff).

We can tentatively confirm that this is not a coincidence: in an equal relationship with another identity, a subject does not tend to choose just one value, just one quality which they consider to be essential for the other's life, but the possibility of value-related attribution arises, and a question emerges for the mysterious otherness in order to understand: "What do you want from me?

3 Conclusions: Crisis and the Revenge of Naturalism in Human-Animal Relations?

To sum up, in in-pack pet food communication the human consumer is portrayed as someone who is predominantly responsible and minorly inclined towards vicarious hedonism, which corresponds to an animal consumer which is predominantly conceived as affectionately dependent, minorly spoilt and slightly independent. The human-animal relation which is prevailingly recognised and appreciated by pet food producers is, without a doubt, one of affectionate and responsible control, in which the identitarian asymmetry between humans and animals is clear. On the contrary, the dialogic relationship, which still appears, is both as radically innovative as it is minor, so minor that we have to ask ourselves how much it can be effectively considered to be a sign of a profound ongoing change in the conception of animality in contemporary western society – which is traditionally dominated by what Descola (2005) denoted as *naturalism.*

Our research stimulates further considerations: are we witnessing the coexistence of different ontologies, in particular the growing diffusion of animism which affirms itself beside naturalism while remaining minor? Or rather, as we suggest, we are witnessing a re-stabilisation of naturalism as dominant ontology in western human-animal relations, a kind of re-entry of animism in naturalism, which allows itself to be disturbed by radically alternative instances (Luhmann 2002), assimilating it (partly neutralising their subversive value) and in this way reaffirming and reinforcing itself.

In this second case, naturalism confirms itself to be the dominant frame for the conception and relations of humans and animals, tempered by animistic instances – as we have already pointed out in an observation of human-nature relations connected to vegetable garden practices (Bartoletti and Cecchelin 2016).

Secondly, we ask if it is possible to find the attribution of an internal continuity between humans and animals (a characteristic trait of animism) when we come to recognise the mystery of an animal's otherness, as we believe we are observing in just one case, but of great interest, that of the communication strategies of Almo Nature. We wonder if the recognition of that mystery is not rather a manifestation of the irreducibility of the animal, an untamed trait, which can never be completely removed from the animal identity – which is always found in the inner tension, which is typically naturalist and western, related to wild/domesticated dualism (Descola 2005).

References

Arluke, A., & Sanders, R. C. (1996). *Regarding animals*. Philadelphia: Temple University Press.

Bartoletti, R., & Cecchelin, G. (2016). Narrazioni e pratiche della natura in città: gli orti urbani tra cortili e foreste domestiche. *Studi Culturali, 1*, 39–60. https://doi.org/10.1405/83211.

Basso Fossali, P. (2008a). *La promozione dei valori. Semiotica della comunicazione e dei consumi*. Milan: Franco Angeli.

Basso Fossali, P. (2008b). *Vissuti di significazione. Temi per una semiotica viva*. Pisa: ETS.

De Marinis, M. (1982). *Semiotica del teatro, L'analisi testuale dello spettacolo*. Milan: Bompiani.

DeMello, M. (2012). *Animals and society: An introduction to human-animal studies*. New York: Columbia University Press.

Descola, P. (2005). *Par-delà nature et culture*. Paris: Gallimard. English Edition: Descola, P. (2013). *Beyond nature and culture*. Chicago: University of Chicago Press.

Descola, P. (2010). *Diversité des natures, Diversité des cultures*. Paris: Bayard.

Eco, U. (1979). *Lector in fabula*. Milan: Bompiani.

Elam, K. (1980). *The semiotics of theatre and drama*. London: Methuen & Co.

Fontanille, J., & Zilberberg, C. (1998). *Tension et signification*. Liegi: Mardaga.

Franklin, A. (1999). *Animals and modern cultures. A sociology of human-animal relations in modernity*. London: Sage.

Franklin, A. (2002). *Nature and social theory*. London: Sage.

Greco, M. (2011). Vademecum del semiologo. In P. Basso Fossali (Ed.), *La semiotica a prova d'esperienza. Sperimentazioni didattiche e metodologie d'analisi* (pp. 349–362). Milan: Unicopli.

Greimas, A. J., & Courtés, J. (Eds.). (1979). *Sémiotique. Dictionnaire raisonné de la théorie du langage* (Vol. I). Paris: Hachette.

Ingold, T. (Ed.). (1988). *What is an animal?* London: Unwin Hyman.

Latour, B. (1999). *Politiques de la nature*. Paris: La Découverte & Syros. English edition: Latour, B. (2004). *Politics of nature. How to bring the sciences into democracy*. Cambridge: Harvard University Press.

Luhmann, N. (2002). *Einführung in die Systemtheorie*. Heidelberg: Carl-Auer. English Edition: Luhmann, N. (2013). *Introduction to systems theory*. Cambridge: Polity Press.

Macnaghten, P., & Urry, J. (1995). Towards a sociology of nature. *Sociology, 29*(2), 203–230. https://doi.org/10.1177/0038038595029002002.

Marrone, G. (2011). *Addio alla natura*. Torino: Einaudi.

Peirce, C. S. (1866–1913). *Collected papers*. Cambridge: Harvard University Press.

Tester, K. (1992). *Animals and society. The humanity of animal right*. London: Routledge.

Williams, R. (1980). Ideas of nature. In *R. Williams problems in materialism and culture* (pp. 67–85). London: Verso.

Cat Cafés and Dog Restaurants

Alice Giannitrapani

Abstract Starting from the assumption that space is a language and that food is also a language, what we will try to show in this paper is how these two languages, together, carry specific concepts of animality. For this reason, we will focus on two different places of food consumption that place animals at the centre of their proposal and that in recent years have spread in urban environments: *cat cafés*, where cats are conceived as subjects to pet, endowed with a predominantly emotional value, and *dog restaurants*, where dogs are in fact the real restaurant clients.

1 Cat Cafés

Cat cafés were introduced at the end of the 1990s in Taiwan and then became very popular in particular in Japan, and from there they spread around the world. These are cafés where you can go for a drink or to eat something while being surrounded by cats that move freely around the premises and are there to be pet and cuddled. Spin-offs of these particular restaurants were various types of pet cafés, some with rabbits, others with reptiles, birds, etc., which in any case are an attempt to establish a blissful connection between men and animals, eager to be with each other, in a relationship of mutual complicity. The creation of such places and their spreading predominantly in Japan appear to originate from a lack of direct contact, a need to relate to animals, which is often difficult to have in Japan, where it is not allowed to keep dogs and cats in most residential buildings and complexes. The narrative programme of connection with felines therefore moves outside of the home, thanks to the creation of cat cafés. The guest here is able to choose how long to stay (paying a time-based fee) and whether to buy food and drink for themselves and/or for the felines (thus becoming the protector/benefactor of the chosen animal).

Most of these places appear to be a true heterotopia (Foucault 1984) where, for a limited period of time, the subject can have an extraordinary experience and put on hold his normal worldly relationships and then return to their usual spaces feeling, perhaps, transformed. The access procedure is orderly and subject to specific rituals

A. Giannitrapani (✉)
University of Palermo, Palermo, Italy
e-mail: alice.giannitrapani@unipa.it

G. Marrone, D. Mangano (eds.), *Semiotics of Animals in Culture*,
Biosemiotics 17, https://doi.org/10.1007/978-3-319-72992-3_7

Fig. 1 Sample of book of cats, a brochure with details on each cat hosted by the café

(another element typical of heterotopic spaces): the customers, when entering, have to remove their shoes and wear slippers, wash their hands and, sometimes, disinfect them. Disinfecting human hands: a somewhat paradoxical gesture that reverses ordinary practices (it is usually considered a good hygienic rule to wash hands after touching an animal). In this case there is a clear projection of categorisations and practices typically cultural on the cat and some kind of sacralisation of the feline: the cat is no longer a potential carrier of germs, but a being that needs to preserve its purity, which must not be contaminated (Douglas 1970). Within these perfectly orchestrated microcosms, well-defined rules exist that can vary from place to place: in some, dogs and children are banned; in others, camera flashes are forbidden, or it is not allowed to offer human food to cats; in some other places, there is even an identification collar that informs on whether the animals have already eaten or are following a particular diet. It's as if these places emphasise the distance between man and animal, though not intended in dysphoric and hierarchical terms but euphorically expressed as a form of respect for the peculiarities of each species.

The cafés also provide *books of cats* (Fig. 1), which contain "records" of the chosen one-night friend: a photo, name, age, possible breed and/or some details about their story. An individualising personification on the one hand gives a totality as a collection (the cats hosted by the café and listed in the booklets) and on the other identifies and tries to create a direct relationship between animal and customer (as to say, you are not playing with "a" cat, but with a "known" cat). In these booklets, the rhetoric of the animal's introduction and of the prefiguration of the relationship with a human being can go in two different directions: on the one hand, the hosted felines can be of pure and rare breeds, and thus focus is placed on the aesthetic impact (beautiful specimens to admire); on the other hand, cats can be rescued strays, and in this way a passionate dimension, made of pity and commiseration but

Fig. 2 Some cat cafés reproduce a natural environment using artificial materials

also of joy and happiness for the happy end that can be created (in fact, these cats are often available for adoption), is emphasised. From this, two different kinds of social discourses generate about these places: animal welfare activist's condemnation of the commercial exploitation of animals (concept of the cat-object) is contrasted by praises based on the awareness campaign carried out in function of and on behalf of animals (concept of the cat-subject).

The architectural solutions adopted are different in every café. On the one hand, there are places that reproduce "nature" but often use materials such as resin and synthetic rubber, vast environments exhibiting a plastic nature and a kind of cartoon scenery where to be playful (fake wood houses, trees, kennel trunks, green areas, Fig. 2). The cat is in these cases part of the environment as a feline; its nature goes back to the origins, to wildness. On the other hand, other places stage a cultural look (classic cafes with tables, chairs, counters) but use natural materials (wood, stone, etc., Fig. 3). Here, cats are considered *urban pets*; we could say they move between the pillows, contemplate the view outside the window and even entertain themselves by watching videos on an iPad (videos that have other cats in the leading role, of course). The cat, therefore, becomes a cultural subject, fully integrated in the urban environment, and a social actor just like a person. These are two solutions that, in their chiasmus-like organisation, reflect two different concepts of animality (Table 1).

In any case, one feature strongly stands out: the *figurative hypertrophy of the animal* or of some of its parts which represent it by synecdoche (typically the paw and the shape of the head). The cat appears on the street sign, on the doors and on

Fig. 3 Other cat cafés look like classic food consumption places where, however, natural materials prevail

Table 1 Spatial organisations (expression plane) and concepts of animality (content plane) in the cat cafés

E	Figurative level	Natural scenery	Cultural scenery
	Plastic level	Artificial materials	Natural materials
C	Conception of the animal	Feline cat	Urban cat

Fig. 4 Figurative hypertrophy of the cat

the cups for human use and is drawn on the walls; fluffy toy cats are scattered around; and the toilet signs have cat/human figures on them (Fig. 4). Living cats and representations of cats refer to each other creating a homogeneous and concentrated thematic universe; everything in the room is aimed at drawing our attention to the feline isotopy, sometimes almost turning it into a caricature. Another recurring fea-

Fig. 5 Valorisation of vertical dimensions in cat cafés

ture is also the utilisation of vertical dimensions: the cat swings from hanging hammocks, shows its abilities crossing suspension bridges and steep pathways and jumps from one plastic tree trunk to the other. The difference in skills, compared to the range of human motion, is therefore actualised, and all that humans can do is, with their nose up in the air, admire the animal's fantastic agility (Fig. 5).

The customer wanders around the venue, buys food to feed the cats, looks and enjoys from a distance the cats ambushing toy mice or tries to play himself but feeds off from a more or less orchestrated passionate side where the drive to be caring is supposed to generate inner relaxation. Feeding is at the same time a pragmatic and an affectionate action, a crossroads that can demonstrate hyper-love for animals as a form of life. Not only cats are there to be pet and nourished (and are therefore docile), people are there to nourish and to pet too (and are therefore tender-hearted). The creation of a cat café is an implementation proposition that transforms both subjects (animals and humans). The restaurant, as a device, contributes to the creation of what Despret (2002) calls "production of existence" (p. 22), highlighting, in this case, the affective dimension.

2 Dog Restaurants

The relationship between the restaurant sector and the canine world is modulated on a series of gradual distances that progressively assimilate (or systematically distance) man and Fido. There are in fact restaurants where dogs are forbidden (in this case the animal, although being considered man's companion, is still considered as a beast), restaurants where dogs are allowed only if specific rules are followed (must have muzzle and leash and in some cases the vaccination booklet), restaurants where dogs can eat while their masters also do (the master is still dominant, but the dog emulates the same human narrative programme) and restaurants dedicated exclusively to dogs (where the border between animality and humanity is neutralised) (Giannitrapani 2017). The dog, on the other hand, is increasingly involved, willing or not, in social practices often related to leisure, typically cultural: food shopping (in some supermarkets there are special trolleys for our four-legged friends), at the beach (there are lidos and deckchairs for dogs), on holiday and in the restaurant. And all these activities reveal, in fact, a certain concept of anthropomorphic animality. In the rooms of the Cipriani Hotel in Venice, for example, wealthy puppies can find a "happy dog corner", an area with hand-embroidered pillows, bowl, custom-made cookies, games, bone-shaped titbits and a tie. Fido can show off his tie at the elegant restaurant where a dedicated dog menu includes, among other dishes, beef tartare at 38 euros or chicken sushi at 29 euros.[1]

The dog restaurant is not simply a pet-friendly place but a culinary place specifically designed for dogs, reproducing all the features of the classic restaurant format: the separation between dining room and kitchen or sometimes the most modern type with an open kitchen, waiters (sometimes in uniform) serving the animals, a sort of table with bowls and daily menus written on the chalkboard (Fig. 6). So, same as a man can more and more go to restaurants and do other things at the same time (buy food or other items, read books, do laundry, etc.; see Giannitrapani 2014), even places dedicated to Fido can be multifunctional (buy takeaway food, take advantage of a grooming service, venture on exercise trails), a kind of dog-world setup but hybrid (dog+master) friendly. And the discourse of the space is consolidated by the discourse about the space, with entire websites that provide information and reviews on dog restaurants,[2] with bone icons replacing the stars, asterisks, forks or goblets typical of restaurant guides and in general of food critics (Marrone 2016).

In this case too, the architectural solutions can go in two directions: dark and ultramodern premises that incorporate the dog in the cultural universe and lighter environments where simple natural-effect materials are used. More often, nature and culture tend to hybridise, showing all their ambiguities: in Qibo, in Rome, the dog finds itself on a synthetic lawn playing in a wooden tunnel (reproducing the park experience), and nearby, there are (simulated) market stalls that offer fresh and/or organic products. In Fig. 7, the dog on the synthetic lawn is practically the extension of the image behind it; it simulates an act of predation on homemade food whose nutritional properties are listed at the back. There is a clear friction between animal perspectives (predation) and human perspectives (knowledge and awareness of nutrition). The conditions of the dog are very ambiguous; it hangs between healthy needs and animal instincts; the status of the dog owner is complex too: on

[1] http://corrieredelveneto.corriere.it/veneto/notizie/cronaca/2016/2-luglio-2016/al-cipriani-menu-cani-tartare-costa-38-euro-240628996970.shtml, http://corrieredelveneto.corriere.it/fotogallery/2016/06/cani_menu/cani-cinque-stelle-240629027413.shtml#2

[2] http://www.bringfido.com/restaurant/

Fig. 6 A dog restaurant in London

Fig. 7 Predation and nutritional values, Qibo (Rome)

the one hand, he respects his animal, but on the other he forges the dog's intentions, needs and desires to reflect his own image and likeness.

2.1 Diet Regimes and Convivial Rituals

Many of these places are innovative in the way they conceive animal food regimes. They value a kind of revival of cooked food and promote the abandonment of the classic kibbles, which, traditionally seen as balanced nutrition, packaged ad hoc for the needs of dog, are now banned as a synonym for rough, industrial food. Changing from dry to wet food and from pre-packaged to fresh, homemade cooking for animals is therefore recommended; but now it's no longer a matter of giving Fido the leftovers of Sunday lunch, but rather offering him the best foods (e.g. organic) or a tailored diet regime that meets his needs.

Restaurateurs say that the real mission of these places is to raise awareness and change dog's diets; the eccentric experience of going out to a dog restaurant is not then set up as a timely practice that ends in one night, but rather as an initial manipulative move and a prerequisite to acquiring a competence (the knowledge about correct animal nutrition that results in the purchase of suitable products) and a durable performance that has broader meaning: the transformation of canine nutrition, which will be permanently embraced by the animal and will be aimed at its physical wellbeing. The new proposed nutritional regime is always enveloped and motivated by a scientific aura (the principles of nutrition extended to the animal world; see Ventura 2014), and the use of a medical isotopy provides validity and credibility to the matter. This is why in almost every dog restaurant there are vet nutritionists available to advise on and recommend a personalised diet for the dog, a new nutrition regime that considers the needs of the man-animal hybrid (balancing dog's needs and master's habits). The standardisation of kibbles therefore opposes the individualised diet, resulting in a change of the dog's status, no longer partitive totality (dog among other dogs), but partitive unity (singular actor with its own physiological and taste peculiarities) (Greimas 1986).

So, Lily's Kitchen, a manufacturer of organic food for dogs and cats, launches a pop-up restaurant in London, open for 6 weeks only, to offer free samples of its product line to users; in Qibo there is a kind of market corner, with small stalls where dishes such as pork and potatoes or hake and rice are displayed, all strictly of Italian origin and homemade. Recurrent lexicalisations used by the restaurant owners, by the vet nutritionists who work there and by the customers are "homemade", "organic" and "correct nutritional regime", terms that, by contrast, refer to industrial food, a type of food sometimes openly referred to as "junk food".[3] Different menus (raw, monoproteic, hypocaloric, vegetarian, etc.),[4] revival of cooked food

[3] http://www.deabyday.tv/cuccioli/whatsnew/guide/7775/Apre-a-Berlino-il-primo-ristorante-per-cani-e-gatti.html

[4] http://www.puntarellarossa.it/2016/06/03/qibo-roma-il-primo-negozio-con-laboratorio-di-cucina-per-cani-e-gatti/

Fig. 8 "Ugly" but healthy and nutritious meals

and display of fresh ingredients whose beneficial properties are indicated do not simply reproduce typically human diets, but connect to an ethical and aesthetic stance on the world, a complex value configuration in which nature, conservation of the environment, love for animals and attention to nutrition (human and animal) converge. In an interview, one of Qibo's customers states that, for example, in addition to cooking for her dog and buying organic products, she has for some time cured it with homoeopathic remedies: the dog is well kept, its master loves it and understands its true necessities, nature is respected and a form of life (not only concerning food) is proudly declaimed. Animal values are based on a man's will, of course, but they are often attributed to the dog itself, according to a process of delegation typical of many discourses on animality[5] ("There is a fully vegetarian menu specially designed for four-legged friends who *wish* [my italics] to exclusively eat fruits, vegetables and cereals", which proudly affirms an article on petsparadise.it).[6] There is, in fact, a gastronomic consciousness made of attention and care for nutrition, based on essentially healthy motives. The served meals, however, often appear to be unattractive, roughly poured into classic animal bowls and sometimes reproduce the shape of the much-reviled kibbles (but this time organic, Fig. 8).

This kind of proposal is contrasted by another line of discourses, spaces and products that instead play on the idea, always human, of hedonistic enjoyment derived from tasting. Food in this case is a sensory pleasure to savour, a beautiful dish to admire (and of course to photograph and post on social networks). The

[5] On this subject see Giannitrapani 2017

[6] http://www.petsparadise.it/altri/cani-e-gatti-buongustai-ecco-i-ristoranti-per-pet-e-padroni/

Fig. 9 "Beautiful" and tasty meals

restaurants therefore offer beautiful and tasty foods and expertly displayed (according to the taste for aesthetics that so many TV cooking shows have taught us to appreciate) and even elaborately designed cakes (Fig. 9). It's as if reproducing the shapes of human food is equivalent to rebuilding the *shape of tastiness*. In this way, the dog not only savours good and tasty thing but rediscovers the taste of sociality, recovering the convivial *côté* of the culinary experience: from which beers for dogs (or wine for cats[7]) and places that offer aperitifs for dogs but also dog birthday parties, themed parties organised by animal party planners[8] and even guerrilla-style events with dog poo-shaped biscuits left around the streets during Halloween, aimed at raising awareness of the hybrid (man + animal) to clean up their dog's poo.[9] It's gastro-mania adapted to animals, the one that, paraphrasing Marrone (2014), we could define as *gastro-zoo-mania.*

[7] http://www.vanityfair.it/lifestyle/animali/14/07/09/ristoranti-e-bar-dog-friendly-aperitivo-con-cane-e-gatto

[8] http://www.kikipelosi.com/category/compleanni-a-4-zampe/

[9] http://www.kikipelosi.com/dolcetto-o-scherzetto-sacchetto/

3 Conclusions

The world of cat cafés and dog restaurants always stages euphoric relationships between pets and humans, leaving aside all those stories of ill treatment, abandonment and conflict that are also present in social discourses about bestiality. There is therefore the wealthy and luxury-loving little dog; the rescued kitty, pampered and luckily saved by another hypothetical sad fate; and the beloved pet unconditionally loved by its master who compares it to a human being and who often proclaims with pride his best essence. And our beloved life companions seem to fit perfectly with the kind of definition that their masters give them, and they seem to enjoy the vegan food they are given, being perfectly at ease within the restaurant devices dedicated to them. Embracing Despret's thesis (2002), we could say that, on the one hand, animals are transformed "according to the habits, the stories and the ways of addressing them" and, on the other, they transform humans. And this is not a causal and unidirectional transformation, but a process in which more or less desirable and more or less random elements are involved. Probably, in our case, gastro-mania (Marrone 2014), the ongoing search for innovation in restaurant formats, the attention to diets, the growing interest in animals, their ever-growing involvement in social practices and discovering their previously unknown needs have helped the creation of dog restaurants, cat cafés, etc. Or, presumably, the focus on human food has turned into attention to animal food, hence the diets that can actually change the tastes and the nutritional system of our pets. The restaurants for and with animals offer us something meaningful, a proposal that aims at the transformation of the human being (increasingly attentive to new and more or less constructed needs of the animal) and of the non-human (increasingly adhering to alternative gastronomic philosophies). And they do it in very different ways. Therefore, to conclude, we will try to summarise the conceptions of canine and feline food that have emerged so far from the various staged (food) ontologies (Descola 2005). On the *animism* position, Descola places all those concepts that postulate between humans and non-humans a continuity at interiority level and a discontinuity at the level of physicality: animal and man are both sentient and sensitive beings, only with different physical, in our case nutritional, needs. Here sits the policy of those dog restaurants that focus on healthy foods and/or on those who feed their four-legged friends with tinned products but of the latest generation (e.g. organic foods). *Naturalism*, although posing a difference on the interiority level, considers man and animal similar on the physicality level; both can therefore eat the same things: it is the old logic of the leftovers or the case of those restaurants that tolerate Fido's presence but don't provide a special menu for him. In the case of *totemism*, man and animal are thought to be equally endowed with a soul and similar on a physicality level and, therefore, characterised by the same nutritional requirements. In this case, the beast and the master almost become one: they adopt the same dietetic regime (i.e. vegetarian) or take part in the same social events (predinner aperitifs, perhaps tasting wine for dogs or beer for cats). This philosophy is rooted in those dog restaurants that showcase the idea of "beautiful" and tasty food or in those places where dogs and men can go to eat together. Finally, with the *analogism*, man and animal are thought to be different, so much in their interiority as in their physicality, and as such they need different foods. It is the logic of the old and traditional kibbles or of those restaurants that ban animals from their premises. But it is also the logic pursued by cat cafés, where the role of the feline (unlike what happens in a dog restaurant) is always distinct from that of man (the cat climbs and does not eat at the table, has a dedicated nutrition system, behaves "as if" it was free). Humans and non-humans are here situated on two completely different levels.

References

Descola, P. H. (2005). *Par-delà nature et culture*. Paris: Gallimard.

Despret, V. (2002). *Quand le loup habitera avec l'agneau*. Paris: Les Empêcheurs de penser en ronde/Seuil.

Douglas, M. (1970). *Purity and danger*. Harmondsworth: Penguin.

Foucault, M. (1984). *Des espaces autres*. Architecture, Mouvement, Continuité 5

Giannitrapani, A. (2014). Ristoranti & co. Identità e comunicazione dei luoghi conviviali. In G. Marrone (Ed.), *Buono da pensare. Cultura e comunicazione del gusto*. Roma: Carocci.

Giannitrapani, A. (2017). *Dal cane-cibo al dog restaurant*. Giochi aspettuali nella dialettica alimentare uomo/animale. Lexia 11.

Greimas, A. J. (1986). Comment définir les indéfinis. *Actes Sémiotiques Doc, VIII*(72), 19–34.

Marrone, G. (2014). *Gastromania*. Milano: Bompiani.

Marrone, G. (2016). *Semiotica del gusto*. Milano: Mimesis.

Ventura, I. (2014). Questa non è una dieta. Linguaggi e strategie del dimagrimento. In G. Marrone (Ed.), *Buono da pensare*. Roma: Cultura e comunicazione del gusto. Carocci.

The Birth of a Pet? The Rabbit

Francesco Mazzucchelli

The rabbit, though it chews the cud, does not have a divided hoof; it is unclean for you.

(Leviticus 11:5)

Abstract In Western cultures, the rabbit holds a double status: it is, at once, livestock and pet. Furthermore, manifold connotations (rabbit can also be hunting quarry, vermin, test animal, etc.) and a rich and long-lasting iconography (it is an iconic animal with a meaningful symbolic presence in many cultures) make this animal a powerful metonymy of the dynamics of the human-animal relation, grounded in incessant renegotiations.

Looking specifically at the Italian cultural context, this chapter aims at exploring the diverse processes of semiotization of the rabbit, and the different (sometimes incompatible) visions of nature lying behind, which may at times take the form of a (only apparently minor) "war of the worlds" (Latour) in which different ontologies come into conflict.

Through a quick analysis of different texts, objects, practices and discourses, I shall underline how various "modes of existence" of this animal confirm the hypothesis of Descola, according to whom different "regimes" of nature (and hence "animality") can coexist (sometimes successfully, sometimes not) in the same society.

F. Mazzucchelli (✉)
University of Bologna, Bologna, Italy
e-mail: francesco.mazzucchelli@unibo.it

G. Marrone, D. Mangano (eds.), *Semiotics of Animals in Culture*,
Biosemiotics 17, https://doi.org/10.1007/978-3-319-72992-3_8

1 A "More Equal" Animal? Rabbit as Frontline of a "War of the Worlds"

War of the worlds[1] may start at dinner, too. An international dinner, say, with rabbit as main course, with Italian, French, British and American guests, whose appetite could be suddenly spoiled by the appearance of a linguistic, ethical, aesthetical dilemma: *Rabbit*? *Coniglio*? *Lapin*? *Bunny*? "How do you cook your *coniglio alla cacciatora*?"; "I used to own a rabbit as *pet* in my garden, when I was a child"; "The main course is *lapin à la moutarde*"; "How can you ask me to eat a *bunny*, they are so cute…"; and so on and so forth; it would be easy to figure out plausible developments of similar conversation-starters that would easily evolve in an escalating battle between incompatible cosmogonies, far beyond an innocent dispute over tastes and distastes.

The example may sound obvious, but its aim is to introduce the main issue of this essay: among all animals, rabbit is probably that one which, more than others, especially in Western cultures, holds this uncomfortable double role – it is "good to eat" but, so to speak, "cute to cuddle" as well.[2] This may probably sound as a too much arbitrary statement, though that could be questioned at least at two different levels. Indeed, firstly, rabbit is not the sole animal which possesses this double status; one of the counterexamples that could easily come to mind is the horse: man's companion and "helpmate" *par excellence*, whose intelligence is widely recognized in the common sense but, at the same time, delicious dish for many others; the horse can turn any discussion into a fight between the defenders of its "quasi human" nature and the fans of *tartare*. Many other examples could be provided, starting from the lamb, whose admission in the paradise of pets seems to have been already announced, and yet, still impersonating the perpetual role of sacrificial victim in Easter (and not only) feasts. A second plausible counter-argument: possibly, if we exclude some minority cases, the rabbit status would be considered less ambiguous if framed within homogeneous cultures. Taking on the opening example, rabbit is considered food mainly within Latin (and especially Mediterranean) cultures, while in Nordic and Anglo-Saxon cultures, rabbit is more and more entering the domestic space as indoor pet, immediately following dogs and cats.[3] Although such objections may sound reasonable, the matter is way more complex, as we will see in the following.

[1] I refer here to the essay by Bruno Latour "Guerre des mondes – offres de paix" (Latour 2000).

[2] As is well known, according to Levi-Strauss (1962), food is "good to eat" if it is regarded as "good to think" as well: its edibility is made possible within a symbolic system which defines also the conditions through which it is allowed to eat some animals, while others are forbidden. Food is then basically a language, expressing a particular structure of the world (see also Marrone 2016).

[3] The distinction between Anglo-Saxon and Latin cultures on this issue should be taken with a grain of salt, as testified by a recent controversy in Venezuela triggered by some statement by the government, which promoted a campaign to consume rabbit's meat in order to boost food availability in Venezuela, urging citizens to stop seeing rabbits as pets (https://goo.g./A8LbPm). See Desmond Morris' observations about indoor pets, that is, those animals who are allowed to live inside the house (and not just on the premises, such as the garden) (Morris 1986).

Indeed, this essay will look at the rabbit (an animal which has already been charged for ages with many connotations and multiple semiotic identities) as at one of the most interesting "frontlines" of the many "wars of worlds" that currently cross our society, whom conflicting factions can count, respectively, on animal rights supporters, chefs, hunters, entrepreneurs, vegetarian consumers, vegans, omnivores and many others, and which through this conflict affects the relationship between men and animals (or human and non-human animals, as someone would prefer). The different thematic (and "ontological") roles which are attributed to rabbits are not, though, a symptom of a mere "clash of civilisation", that is, a collision between different "common senses" in different cultures,[4] but rather of a cross-cutting conflict irrupting inside society, producing axiological (ontological?) fractures with diverse visions of "world composition" confronting each other. Then, the hypothesis of this essay is that if we should find an allegoric emblem, a metonymy, of the manifold ways to express the man/animal relationship, the rabbit may probably be deemed as the most appropriate candidate. For many people rabbit is a livestock animal, main "character" of many typical dishes of numerous culinary traditions, while for many others it's just a cute pet and a beloved tamed animal: many (and often incompatible) visions of nature and culture are developed around the "rabbit body". In the following, through a quick analysis (especially referred to the Italian cultural context) of some texts, discourses, objects and practices in which the rabbit plays the main role (juridical texts, radio shows, animal accessories, recipes, etc.), I shall show how the coexistence/concurrency of different "modes of existence" of a discursive subject/object corroborates the argument according to which naturalistic, totemistic, animistic and analogistic regimes (Descola 2005) can cohabit in current societies, sometimes composing each other, sometimes not (Marrone 2011, 2012; Latour 1999).

2 The Many Dresses of the Rabbit: A Polysemic Iconology

Classified into the order of *lagomorphs* (but it was considered as belonging to *rodents* until 1912) and into the family of *leporids*, herbivores comprising eight *genera* (*European rabbit Oryctolagus cuniculus* and *Sylvilagus* are the most popular, known respectively as European and cottontail rabbit, but the *genus* includes also the family of *Lepus*, whose members are commonly known as hares and jackrabbits), the rabbit is domesticated relatively late, in the sixteenth century. Few animals can be compared to it for the capacity of assuming, in the same culture, a range of so variable and self-contradictory valorizations. It is, as said, *livestock*, a source

[4] In his essay "War of world, offers of peace" (2000), Bruno Latour comments the notion of "clash of civilization" by Samuel Huntington. This notion, Latour says, has the advantage to focus on conflict, but it is still affected by classic culturalistic positions and then fails in "recognizing the true frontlines of such clash, that do not cross in any way the odd aggregates designated by him" (ibidem, *my translation*).

of food and delicacies, but also of fur and leather, and quarry as well, favourite game for many hunters. It is also a *pet* and companion animal, a recipient of human affectivity, not only in recent times but, at least in Anglo-Saxon cultures, in a longer tradition (although its repositioning from outdoor to indoor pet should be better investigated). Occasionally rabbits can also be *food for other pets*: it seems indeed that the smaller specimens are sometimes fed to domestic snakes and other carnivore pets. It can be also used as *test animal* in pharmaceutical and scientific laboratories. Finally, sometimes rabbits are also regarded as *environmental plagues*, capable of wasting harvests, and then threat as well to be contained and eradicated (selective viruses that target exclusively rabbits have been synthetized in labs).

Rabbit is then capable of summoning an abundant set of narrative functions, thematic roles, tropes, topics and discursive configurations: it is at the same time subject and object in different practices – dietary, affective, sport (as in hunting), scientific practices and so forth. Moreover, this narrative polysemy corresponds to a parallel "iconographic polysemy", which enriches the array of narrative roles: this animal can be found in many cultures, mythologies and cosmogonic representations but also in arts and literature, fairy tales, films and cartoons. A complete historical and cultural survey of the different significance of rabbit in different cultures falls outside the aims of this paper, although some quick example could introduce the complexity of the multiple semiotic identities of this animal.[5] Prey *par excellence*, the rabbit is, on the one hand, a metonymy for timidity and cowardice (in some languages it is used as an insult to indicate a weak person), but for tenderness as well; on the other hand, it has always personified also fertility, intended both as sexual fecundity and lust (e.g. Easter bunnies, linked to a symbolism of fertility, or, for a more trivial example, the bunny silhouette used as logo in the *Playboy* magazine). There is a long-standing similar iconography, well established across centuries and different cultures: representations of rabbits and hares can be found in ancient Egypt (when the god Osiris was often depicted as a rabbit), in Aztec and Mayan cultures, in the Greek-Roman world and in the Christian tradition (where this animal can represent both purity and lust). However, this iconography is counterbalanced by a specular one, more "dysphoric", in which all the traits of cowardice, tenderness and purity are reversed in their contrary or contradictory terms, into a symbolism that inverts the other one (even though such contradictory features seem to coexist in some, especially folkloric and "exoteric", discursive genres). During the Middle Ages, for instance, representations of rabbits can be found in many miniature codices as devilish figures, and sculptures of demon rabbits are sometimes present in religious architecture (one famous example is the rabbit contained in the façade of the Chartres Cathedral in France) (Fig. 1).

A complete list of all the literary and cinematographic references to rabbits would be impossible to complete in these few pages; it will suffice here to mention just some well-known examples, such as the black rabbits imagined by Carlo Collodi, who carry on their shoulders the coffin that announces the coming of death to Pinocchio, or the "white rabbit" by Lewis Carroll, or, again, the many rabbits

[5] See, among other works, Davis and Demello 2003; Baker 1994.

Fig. 1 Miniature inside a manuscript from the fourteenth century ("Smithfield Decretals"), depicting a rabbit which decapitates a man

Fig. 2 Frame from Rabbits, series of short movies by David Lynch (2002)

featured in cartoons (Bugs Bunny, Roger Rabbit, etc.), all of them personifying either an almost wicked smartness or a lucky clumsiness. Notable rabbits appear in famous movies, from the wacky killer rabbit by *Monty Python*, who despite his harmless look is a merciless exterminator, to the *Rabbits* by David Lynch (probably inspired by the rabbit heads featured in "My American Uncle" by Alain Resnais), from the rabbit-masked ghost in *Donnie Darko* to the rabbit kid who seems to live in a nonsensical world in the movie *Gummo*. It would be easy to extend the list, as rabbits still keep representing a strong figure in the pop imagery as well (Fig. 2).

The rabbit is then an ambivalent, "uncanny" character, seldom carrying values of disquietude and obscure meanings: a recurrent cultural trope, a sort of *zoeme*,[6] to use the notion by Levi-Strauss to designate the semantic functions covered by some animal species (Levi-Strauss 1971).[7] According to Levi-Strauss, who stresses the ambivalent nature of *zoeme*, "it appears that some animals are more suitable than others in embodying such a function, both because of a particularly significance of their constitution or behaviour, or because, still for natural disposition, the human thinking is able to catch faster and easier the features of a certain type" (*ivi*: 525 trad. it., *my translation*). Hence, it cannot be excluded that a sort of "zoematic capacity" of the rabbit shall be attributed to some salient elements of its figure, such as the mimetic ability that contrasts with its morphologic complexion (i.e. the ear shape, an "hypercodified trait" in many cultural representations) or the fact of being a dumb animal (oddly except when it dies, when it emits a feeble kind of call).

3 A Threshold Animal? Rabbit as *Cuniculus* Among Ontological Domains

According to a popular classification, animals are divided among those we eat, those with whom we watch TV and those we fear; the general impression after the previous pages is that the rabbit could easily fall in each of these positions: it is "good to eat", it can be a good companion animal as well, but we can also be afraid of it, as testified by some long-lasting "obscure iconography" circulating in many semiospheres. In light of the above, one could then suggest that the rabbit, with its multiple connotations, is capable of undermining many cultural systems of classification. The contradictory thematic roles that it can play transform it into a sort of threshold animal between different worlds, similarly to the White Rabbit by Lewis Carroll: rabbit is a trickster, a (potentially deceiver) mediator.[8] In other words, it represents a point of overlap, and of division at the same time, between visions of culture and nature that are deeply and structurally diverse, between different ideas of the human-animal relation (and of the same definition of animal): a *cuniculus* (word for both rabbit and tunnel in Latin) in case one wanted to try a creative use of the etymology of the Latin world. In the next pages, I will look at the rabbit as at an "exchange device", a sort of "discursive switch", capable of acting (in myths, tales,

[6] I owe the suggestion of the concept of *zoema* to Paolo Fabbri.

[7] "One could say that myths use what we could name zoems, that can be reduced to bundles of variously combined differential elements, similarly to phonemes for the linguists" (Levi-Strauss 1971: 69 my translation).

[8] See Baker 1994. Greimas instead defines the *trickster* as an actantial role that can be defined on the plane of veridiction and that "has the function of modifying the knowledge mechanism and make ineffective his interpretative action, producing the cognitive space of illusion" (Greimas 1976: 74–75 my translation). The *trickster* is then "someone who deceives himself for someone else" (ibidem).

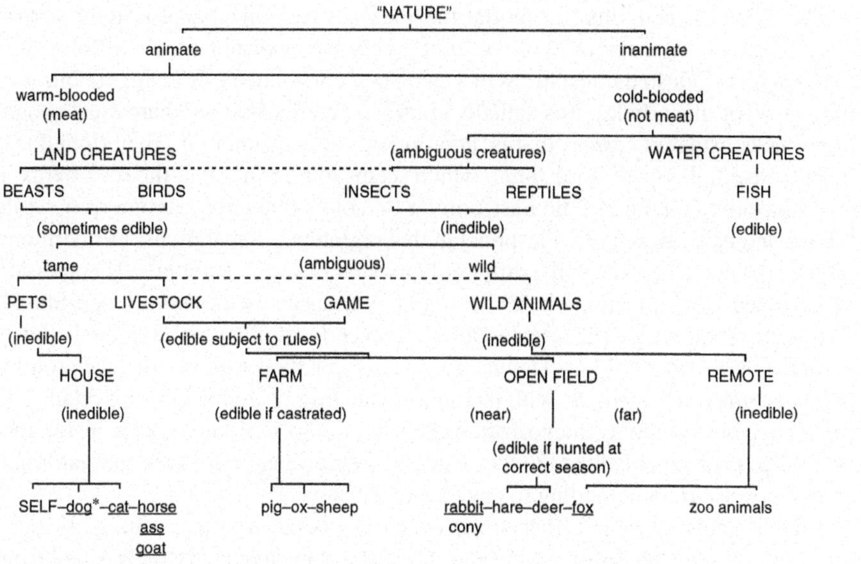

*The species underlined on the bottom line are those which appear to be specially loaded with taboo values, as indicated by their use in obscenity and abuse or by metaphysical associations or by the intrusion of euphemism.

Fig. 3 Classification proposed by Edmund Leach

narratives) as an *isotopic connector/disconnector*: one of the many "figures of animality" (although not the only one) which can function not only as a "translator agent" between humanity and animality but above all among discursive domains (which should be considered "ontological chunks").

In order to investigate the vagueness of any classification encompassing the rabbit, a valid starting point could be the categorization proposed by the anthropologist Edmund Leach in a survey about taboos, language and animal categories (Leach 1989) (Fig. 3).

The classification advanced by Leach is interesting to the extent it suggests – beyond a differentiation on animal statuses depending on their relations with men (pet, livestock, game, wild animals) – a further classing related to places (home, farm, field and wood, distant places). It is not relevant that in this classification the rabbit is never a companion animal, while it is worth noting how, in this classification too (which is referred basically to the English culture), the rabbit holds again an ambivalent and vague position: not completely tamed but neither entirely wild, the rabbit lies in an in-between space, between nature and culture, domestic and savage space: "the obscene rabbit, which is ambiguously game or vermin, occupies an intermediate status between the farm and field categories" (Leach *op. cit.*: 161).[9]

[9] Remarkable are also the observation by Leach about the various denominations of the rabbit, for which the same word is used to designate both the animal and the meat destined to consumption (similarly to other small animals). Moreover, the word "rabbit" is a substitute for a taboo word,

On the other hand, this "categorial fuzziness" is recognizable also in the Jewish doctrine about the prohibition of consumption of the unclean animals' meat: if it is true that many attested translations of *Leviticus* seem not to allow eating rabbits (see the *exergo* of this article), it is actually controversial if rabbit and hare meat should or not be considered *kashèr*, that is, appropriate for consumption. The original text in Hebraic employs the word *šafan*, which seems to refer more to the rock hyrax, a common animal in the Middle East, similar to rabbits (but not a Lagomorph), while rabbits and hares were not widespread in that region in biblical times. Furthermore, rabbit does not "chew the cud", differently to what is said in *Leviticus*. The question is discussed also in Islam, where there is no consensus on the *halal* ("permitted") quality of rabbit meat. The unavoidable reference is, at this point, the book *Purity and Danger* (1966) by Mary Douglas, who analyses the system of dietary prohibitions contained in *Leviticus*, with the aim of studying its significance in relation to the systems of symbolic categorization, coming to the well-known conclusion that prohibited meat can be linked to the "categorial ambivalence" of their animal,[10] that is, to their uncertain belonging to any shared category.[11]

Rabbit status is then further complicated by its even greater ambiguity (also reflected in Linnaean taxonomy, where rabbits have been recently reclassified from *Rodentia* to *Lagomorpha*, as briefly discussed in par. 2): if we look at this ambivalence through Leach's classification, in which edibility is put in relation with prohibition/taboo, rabbits seem to fit at once (not only in different cultures but sometimes in the same cultural system) in the three classes – edible and consumed, edible but tabooed and edible but not recognized as food (hence being included among the substances that are unconsciously tabooed, according to Leach's approach).

4 Accessories for Rabbits

In order to verify empirically the still persistent categorial fuzziness of the rabbit, an easy exercise would be enough: a rapid visit to one of the most popular marketplaces of our times (and it is well known how marketplaces are forced to follow accepted categorization systems), that is, the virtual marketplace of *eBay*. If one launched a

which testifies, on the one hand, the pre-existent quasi-taboo feature of this animal (that used to be called "coney" or "cunny", words that are very similar to "cunt" and from which derives the childish "bunny") and, on the other hand, the attempt of a "linguistic domestication" (Leach op. cit.: 160). In another excerpt Leach insists on the ambivalent status of rabbit, regarded at the same time as plague and food (but only in some moments of the year): "Although vermin and pests are intrinsically inedible, rabbits and pigeons, which are pests when they attack crops, may also be classed as game and then become edible" (Leach *op. cit.*: 159).

[10] For instance, the pig does not fall into the shared classificatory scheme because it has a cloven hoof as the ungulates, but it is not ruminant.

[11] This conclusion is compatible to Leach's comments, when he says that "Language gives us the names to distinguish the things; taboo inhibits the recognition of those parts of the continuum which separate the things" (Leach, op. cit.: 155).

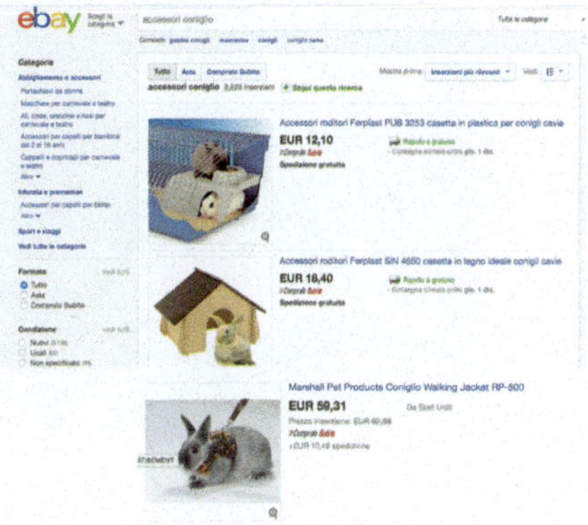

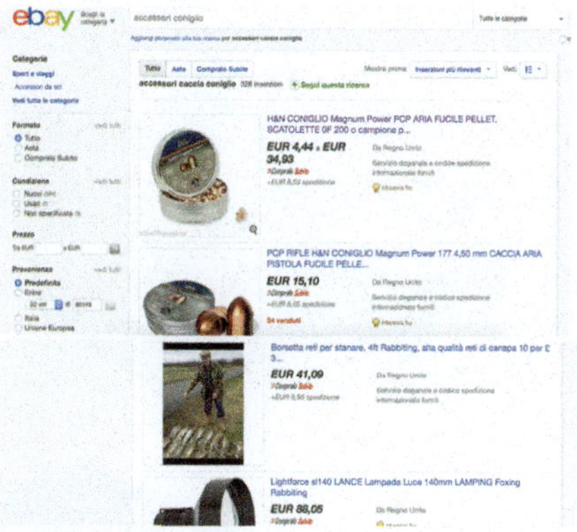

Figs. 4 and 5 "Accessories for rabbits" on sale on Italian eBay

simple research, typing "accessories for rabbits" inside the search box of the internal *eBay* search engine, a miscellaneous and odd list would appear, bringing together, in the very same page, feed boxes and animal feed, hunting trap cages and kennels, play-sets for pets and bullets for the hunting.[12] The heterogeneity of the list would be better understood once one comprehends that all the items belong basically to two mutually exclusive classes, depending on the narrative programmes they activate[13]: on the one hand, *care-oriented accessories* (kennels, feed, toys, brushes, etc.) and, on the other hand, *seizure/death-oriented accessories* (bullets, traps, blinding lamps, etc.), or in any case object destined for an utilization of the rabbit that involves the killing of the rabbit (I did not examined carefully the research list, but I imagine that it would be plausible to find at some point cooking, or even butchery, tools) (Figs. 4 and 5).[14]

This typology of objects both identifies two opposed classes of narrative programmes and two corresponding semantic areas as well (two isotopic but also axiological areas), where many "discourses", texts, objects and practices related to rabbits find their place. On the one hand, we have the semantic isotopies of "rabbit as pet" (related to narrative programmes of care), with such themes as "animal inclusion in human sphere of affectivity" but also "extension of animal rights" (being those two isotopies not necessarily equivalent). On the other hand, we find other isotopic classes and discursive configurations, for instance those related to discourses about domestication, hunting, nutrition and cooking art, but about scientific research too (discourses about vivisection or research on animals).

At a first glance, which shall be later corrected, the divergence between these two classes (*care* vs. *utilization*) could seem correlated to two different axiologies: one that *subjectifies*, that is, produces a subjectivity for the rabbit (hence becoming a

[12] Interestingly enough, results may vary significantly in different regional version of eBay. The examples shown in figures are from Italian *eBay*.

[13] For the notion of narrative programme (shortened: PN) and for all the other categories of narrative semiotics, see Greimas and Courtés 1979, *ad vocem*.

[14] In Italian *eBay*, for both feed and bullets, the same word, "pellet", is often used.

Fig. 6 Different models of cages and kennels for domestic rabbits

companion animal, a pet, a living being worth of protection), versus another one that *objectifies* or, in other words, discursively constructs the rabbit as an object, subjected to specific narrative programmes (rabbit as prey to be seized, as food to be cooked, as plague to be eradicated, etc.)

Following this line of thought, one could think that what is at stake here is an "animalist axiology" versus a "naturalistic" one (to draw on Descola's terminology), but things seem to be more complex, though. First, beyond the example previously given about rabbit accessories, "discourses of care" do not necessarily exclude "putting to death discourses"[15]: the farmer, for example, can breed his animals sometimes taking care of the animals up to the moment of the slaughter, also with a kind of emotional concern. Similarly, the "discourses of care" (and of inclusion of the animal in the domestic sphere) may vary remarkably, so as a rabbit can be pet in many different ways.

In Fig. 6, for example, three different cages for pet rabbits are seen: without any need of a deeper analysis, one can easily appreciate to what extent each of these objects designs a different (more or less symmetrical) man-animal relation, different narrative functions for both the (human and animal) actors (with dissimilar valorizations, modal investments, narrative programmes), but above all different manners of "being animal" (and an "animal owner").

[15] I want to thank in particular Francesco Mangiapane whose observations and comments made me rethink this passage.

5 Axiologies of Care vs. Axiologies of Utilization

Manifold regimes of "animality" (and of posing the man-animal relation) can then cohabit both in "discourses of care" and in "discourses of utilization". For instance, if we look at some examples of discourses that can be referred to the "axiology of care", we would find that many of them hesitate between dissimilar modalizations of "being animal" of the rabbit. Let's consider some texts that are ascribable to an axiology of care, such as some online "handbook" targeted to a person who decided to adopt a rabbit inside the domestic space[16]: many of these texts insist on descriptions of rabbit "personality", of its habits and needs (which may vary depending on the race, but nevertheless with some constants: intelligence, affectivity, necessity of specific attentions, etc.). What comes out from these depictions is essentially an "interaction regime" between the animal and its owner, with various degrees of asymmetry (comparable to the diverse types of interaction drawn by the cages/kennels seen in the previous paragraph) that can range from the "operation" (animal as object) to the "manipulation" (animal as subject),[17] in which the owner plays almost always the narrative role of "Sender" (in French *Destinateur*, in the terminology of Greimas and Courtes).[18] More seldom, an equal relation, between subjects, is thematized, similarly to what can be found in some comparable handbooks targeted at dog and cat owners (but a deeper analysis would be necessary to confirm this hypothesis).

But the most interesting texts about rabbits and their position inside the "human-animal collective" are perhaps the recent petitions and legislative proposals oriented towards the acknowledge of particular legal protections for this animal. In the Italian context, where these preliminary observations have been primarily conducted, the most important among these texts is surely the recent petition named "Coraggio coniglio" ("Come on Rabbit") launched by LAV (*Lega Anti Vivisezione*, Anti-Vivisection League), which later has inspired a draft law initially proposed by the Democratic Party (PD) in 2015 and then resubmitted in 2016 by the People of Freedom Party (PdL). LAV's petition denounced the cruel suffering of the livestock rabbits and aimed at collecting signatures for a legislative proposal which, exerting a leverage on the growing presence of rabbits in the Italian houses as an "affection animal", would acknowledge the status of "family pet" of the rabbit (as for cats and dogs) and ban the rearing of rabbit for consumption purposes. The two draft laws are inspired by the same spirit of the petition, asking for the banning of breeding, slaughtering and consumption (with terms of imprisonment for violators). Both the petition and the draft laws move from what we called the double status of rabbit: "cute to take care for but good to eat", according to LAV's claim, and new "affection animal", according to the draft laws; for both texts the killing of rabbits cannot be

[16] All the examples considered here refer, again, to the Italian context. See for example the AEE information sheet at the address http://www.aaeconigli.it/schede.php?id=1

[17] Cfr. Greimas, Courtés 1979, *ad vocem*

[18] See, again, see Greimas, Courtés 1979, *ad vocem*

tolerated any more. The "new collective sentiment" (this is the phrasing used in the draft law) towards this animal urges a different juridical protection of its rights, however in contrast with engrained dietary habits (in Italy there are more than 8000 intensive breeding farms, and the 25% of the European production is Italian).[19] The incompatibility of the two status, livestock and pet, is ratified by a presumed change in a "collective interpretative habit", influencing more and more people. In both cases, the main argument of these initiatives oscillates between the acknowledgment of the necessity of an extension of specific rights to rabbits inasmuch as sentient being with a privileged relation with humans, and an extension of legal protection of human affectivity, "offended" by the violence perpetrated to a "now more friend" animal.

While the discourses of rabbit care (as affection animal) have to do basically with "existential values", the discourses of utilization concern the transformation of the rabbit in a "usable object" (and should be considered "utilitarian values"). Texts of this last type vary a lot, but they all have in common the fact that they frame the rabbit inside a chain of operations, including their killing, oriented towards the transformation of their bodies into something else, e.g. eatables. Thus, this second class of texts includes recipes, tutorials for the treatment of fur, guides for hunters, but also handbooks and guidelines for breeding. Differently from the subjectification employed in discourses of care, these texts operate an objectification and a use (a semiotic process of construction of an object of value, to use the wording by Bastide 2001). This can result in some "semiotic short-circuit effects": whereas, in the subjectifying discourses of care (aimed at establishing the rabbit as a subject), the animal appears worth of protection only inasmuch as recipient of a human affectivity (so that the human remains the only true juridical subject), in the discourse of utilization the objectifying and naturalistic style summons an axiology of care which somehow acknowledges some subjective rights to the animal (it would suffice to think of the recent notion of "animal well-being" that nevertheless revokes old praxis of caring and respecting the animals destined to slaughterhouse). Whereas the axiology of the rabbit as animal of affection insists on the incompatibility of its double status, that of the rabbit as livestock seems instead to resolve this contradiction sometimes including care into the narrative programme of "putting to death", in an acceptation that, paraphrasing Hjelmslev, one could define "participative": the narrative programme of putting to death foresees somehow narrative programmes of care. Despite appearances, in both cases, both discourses can be regarded as "naturalistic", to use Descola's notion, inasmuch as the continuity of physicality and discontinuity of interiority is saved, posing an emphasis either on the first or on the second term; for instance, even in the case of the LAV petition, where the rabbit is presented as man's companion with an inner sensitivity (a position that could be ascribed in principle to an animist conception, to the extent it insists on the similarity of the interiority of men and animals), the adopted register is a naturalist one, probably to make the argument more compelling.

[19] Figures from Italian Ministry of Health, available at the official website.

Fig. 7 From the official blog of the Italian TV Show "La Zanzara"

This quick overview on some possible typologies of "discourses on rabbits" gives a small glimpse on the complex dynamics of semiotic "explosions" inside the semiosphere (Lotman), generated by the copresence of contradictory narratives, themes and values concerning the semiotic status of this animal, such as the vegan ideology, animalism, hunting, vivisection and all the other "discursive nodes" of the intricate and self-contradictory net of discursive practices which crosses today the "body of the rabbit". The relation of contradiction between discourses of care and of utilization is not limited to a mere alternation between mode of "affirmation" and "neutralization" of the copresence of these two "value poles"; actually, the most interesting cases are those that could be named "diplomatic incidents"[20] between the two worlds. One example of provoked "diplomatic incident" related to our topic can be found in one episode of the Italian radio show "La Zanzara" (the mosquito), broadcasted in an important Italian radio station. The host, Giuseppe Cruciani, disagreeing with the above-mentioned legislative proposal about rabbit, stages a sort of anti-vegan performance, during the live show, consisting in bringing a dead rabbit in the radio studio, asking a cook to prepare a dish out of it and eating it (Fig. 7).[21]

[20] I am referring here especially to the notion of "diplomat" adopted by Paolo Fabbri and taken from Isabelle Stengers: "the diplomat is someone who is obliged to deal with two worlds, one the one side, the world of the scientific studies and, on the other side, the natural world" (Fabbri 2012: 159, my translation).

[21] Some excerpts from the show are available here (in Italian) https://goo.gl/rVvV8J

A similar happening took place during another radio show, this time in Denmark, but with some meaningful differences: in this case, the host kills during the live show a rabbit (right under the eyes of an animal rights activist with whom he was arguing about this topic). Here, the aim was to expose the hypocrisy of meat consumers, who, to use the words of the host, "pretend they do not know that, anytime they eat meat, an animal dies",[22] condemning the fact that all the concerns about animal well-being are limited only to some animals, those capable of triggering "some human affects".

These are obviously two very dissimilar examples, entailing different axiologies: in the case of the radio show "La Zanzara", the staging of the moment of slaughtering (and consumption) can be interpreted as a provocation, aimed at claiming the legitimacy and "normality" of meat consumption in front of who wants to deny and prohibit that; in the case of the Danish radio show, the "killing performance" points at making visible what is usually invisible, that is, that process of objectification mentioned before which turns the animal into meat and food. Both cases display an operation of "re-enunciation" of two narrative programmes belonging to the two classes mentioned earlier (putting to death and utilization), yet charged of different thematizations: in one case, the legitimation of the alimentary consumption of rabbit meat and, in the other case, the contradictory coexistence of different axiologies.

6 A Narrative Zoosemiotics?

This quick, and somehow fragmentary, *excursus* around the rabbit figure and its diverse narrativizations (and the different regimes of compatibility of such narratives) does not claim to offer neither an exhaustive picture nor a deep analysis on this topic, but rather aims at presenting a proposal in view of a necessary in depth exploration on a field which is surely more articulated and concerns the general topic of the semiotization of the animal, of the man-animal relation and of the related "politics of nature" (Latour 1999; Marrone 2011, 2012). The choice of rabbit as object of investigation has to do with the idea that analysis on "figures of animality" (and then on the structures and processes of semiotization of the animals) can cast a light on how the human-animal relation is constantly rebuilt and renegotiated.

But what is the specificity of a semiotic approach to the human-animal relation question? In *The Open*, Giorgio Agamben (2002) applies to this issue the model he had already proposed in his book *Homo Sacer*, in which he distinguishes between two kinds of life: *zoé*, the "bare life", that is, the biological principle belonging to all human beings, and *bìos*, a "form of life", that is, the social and cultural ways of life. In other words, for Agamben, man is not a given and finished starting point, "not a substance, nor a clearly defined species: it is, rather, a machine or a device to produce the recognition of the human" (Agamben 2002: 36, *my translation*). Hence, man is the result of a "dispositif", a "machine", that determines an "inclusive exclusion"

[22] See https://goo.gl/ta4k3j

through which the human is produced by the exclusion of what, in different historical stages, is defined as "animal". Humanity (and hence animality) would then been constructed time after time through different dynamics of (exclusive) inclusion and (inclusive) exclusion induced by the "anthropogenic machine": thus, if the "modern machine" animalizes the human (the not-yet human), isolating the non-human in the man, the "ancient anthropogenic machine" is used to perform this separation through the humanization of the animal (e.g. the slave as animal similar to men). This model is rethought by two young Italian philosophers, Caffo and Sferrazza Papa (2015), and criticized for its "weak anthropocentrism" (Caffo and Sferrazza Papa 2015: 137).[23] Caffo and Sferrazza Papa consider that the two modes of the life (*zoé* and *bìos*) are indeed "animality" itself and then propose a substitution of the anthropologic machine with a "cascade scheme": "we define animality as a natural object (but, as we will see, it is possible to think it as a social object as well) that constitutes the ground, or rather, the bare fact or, in other words, the starting inscription, of two other kinds of, obviously social, objects, which are humanity and citizenship" (Caffo, Sferrazza Papa 2015: 132, *my translation*). Despite the interest and originality of the arguments proposed by Caffo and Sferrazza Papa, I want to question the suggestion of considering animality as a "natural object" (even if mitigated by their clarification about the possibility to consider it also under a social dimension), which complies with the ontological theory outlined by Maurizio Ferraris (and his distinction between social and natural objects; see Ferraris 2009).

The peculiarity of a semiotic approach to the question of the human-animal relation probably lies on the dissent on a theoretical position which considers animality as a "mere fact" or "starting inscription", in other words a "natural object". Rather than on an attestation of the ontological status of such a term as animality, the problem would instead lie on the condition of accountability of the discursive practices that bring towards the stabilization of an ontology or towards the "competition" among diverse ontologies, too. According to this theoretical attitude and gaze, in the previous pages, I aimed at establishing the discursive status of two contrasting "local ontologies" concerning precisely the different "citizenships" of the rabbit in its relations with man: different "modes of existence", to use Latour's words (2012), that clarify the epistemological rather than ontological features of the considered subjects/objects. A similar approach does not confuse epistemology and ontology, but it is rather interested to the diverse "senses" of animality and to the narrative and discursive configurations through which the human-animal relation is represented and produced.

From this angle, maybe also the rabbit – this "*cuniculus* animal" roaming among manifold ontologies, this "diplomat" (Fabbri 2012) between different modes of being an animal, different contrasting visions of the world but also, more simply, between different "discursive planes" – may humbly indicate some meaningful aspects of the complexity of the multi-natural fabric (Marrone 2011) that constitutes our plural systems of values, and their possibilities of reciprocal translations, of coexistence and inconsistence.

[23] About the application of Agamben's model to animal rights; see Filippi 2012.

References

Agamben, G. (2002). *L'aperto. L'uomo e l'animale*. Torino: Bollati Boringhieri. English edition: Agamben, G. (2004). *The open: Man and animal*. Palo Alto: Stanford University Press.

Baker, M. P. (1994). The rabbit as trickster. *The Journal of Popular Culture, 28*, 149–158.

Bastide, F. (2001). *Una notte con Saturno. Scritti semiotici sul discorso scientifico*. Roma: Meltemi.

Caffo, L., & Sferrazza Papa, E. (2015). Nuda vita come animalità. Un argomento di ontologia sociale contro Giorgio Agamben. *Lo Sguardo – rivista di filosofia, 18*(II), 129–138.

Davis, S. E., & Demello, M. (2003). *Stories rabbits tell. A natural and cultural history of a misunderstood creature*. New York: Lantern Book.

Descola, P. (2005). *Par-delà nature et culture*. Paris: Gallimard. English edition: Descola, P. (2013). *Beyond nature and culture*. Chicago: University of Chicago Press.

Douglas, M. (1966). *Purity and danger: An analysis of concepts of pollution and taboo*. London: Routledge.

Fabbri, P. (2012). Natura, naturalismo, ontologia: In che senso? Conversazione con Gianfranco Marrone. In P. Fabbri (Ed.), *L'efficacia semiotica. Risposte e repliche*. Milano: Mimesis.

Ferraris, M. (2009). *Documentalità. Perché è necessario lasciar tracce*. Roma: Laterza. English edition: Ferraris, M. (2012). *Documentality: Why it is necessary to leave traces*. New York: Fordham University Press.

Filippi, M. (2012). Menti animali. In S. Castignone & L. Lombardi Vallauri (Eds.), *Trattato di biodiritto. La questione animale*. Milano: Giuffré Editore.

Greimas, A. J. (1976). *Maupassant. La sémiotique du texte : Exercises practices*. Paris: Editions du Seuil.

Greimas, A. J., & Courtés, J. (1979). *Sémiotique. Dictionnaire raisonné de le théorie du langage*. Paris: Hachette. English edition: Greimas, A. J., & Courtés, J. (1982) *Semiotics and language: An analytical dictionary*. Bloomington: Indiana University Press.

Latour, B. (1999). *Politiques de la nature. Comment faire entrer les sciences en démocracie*. Paris: Éditions La Découverte et Syros.

Latour, B. (2000). *Guerre des mondes – offres de paix. Ethnopsy, les mondes contemporains de la guérison 4* (Numéro spécial, Colloque de Cerisy, Guerre et paix des cultures) (pp. 61–80)

Latour, B. (2012). *Enquête sur les modes d'existence. Une anthropologie des Modernes*. Paris: La Decouverte.

Leach, E. (1989). Anthropological aspects of language: Animal categories and verbal abuse. *Anthrozoös: A Multidisciplinary Journal of the Interactions of People and Animals, 2–3*, 151–165.

Lévi-Strauss, C. (1962). *Le Totémisme aujourd'hui*. Paris: PUF.

Lévi-Strauss, C. (1971). *Mythologiques IV. L'homme nu*. Paris: Plon.

Marrone, G. (2011). *Addio alla natura*. Torino: Einaudi.

Marrone, G. (2012). *Semiotica della natura. Internaturalità e significazione*. Milano: Mimesis Edizioni.

Marrone, G. (2016). *Semiotica del gusto. Linguaggi della cucina, del cibo, della tavola*. Milano: Mimesis Edizioni.

Morris, D. (1986). *Dogwatching*. New York: Three Rivers Press.

Part II
Animals in the Texts, Texts as Animals

Bestiality: Animal Cultures

Gianfranco Marrone

It is possible to read Kafka's animal stories for quite some time without realising he isn't talking of men When one is suddenly confronted with the name of the creature -the ape, the dog or the mole – one raises one's terrified eyes and realises to already be far from the human continent.

Walter Benjamin

Abstract Multinaturalism shouldn't be viewed solely as the paradigmatic existence of multiple natures in different cultures – or "ontologies" as Descola refers to them – but also and most importantly as the syntagmatic presence, and the resulting contrast, of various ontologies within the same culture. From a semiotic point of view, then, the so-called ontologies are considered to be *effects of meaning* derived from precise *discursive dispositifs*. Describing such dispositifs is the aim of this paper.

1 Starting from the Middle

The title of this article – *bestiality* – can be read in two ways: as the theme of the discourse (I am discussing bestiality) and as a value of enunciation (I say bestiality). In the latter, we understand the term "bestiality" in the French sense of *bêtise*, stupidity, that famous idiocy that fascinated many writers (Montaigne, Flaubert, Barthes, etc.),[1] but also, in a not entirely oppositional sense, "wild animality", being a beast in the sense that one lacks civil ways of being and doesn't possess adequate education and sufficient culture. Essentially, behaving like a beast but also being

[1] See Marrone (2012a).

G. Marrone (✉)
University of Palermo, Palermo, Italy
e-mail: gianfranco.marrone@unipa.it

G. Marrone, D. Mangano (eds.), *Semiotics of Animals in Culture*,
Biosemiotics 17, https://doi.org/10.1007/978-3-319-72992-3_9

reduced to being just like a beast.[2] The article's subtitle – *Animal Cultures* – sets a more precise isotopy: that of the nature/culture relationship and its oxymoronic crossover towards a kind of "cultural nature" or – though this is not the same thing – a "natural culture". This semantic restriction, however, does not entirely cancel out the previous series of themes and their possible metaphorisations, which I would like to keep in the ethical background of this piece.

The space of the discourse in which I wish to place myself is that produced around the problematic notion of *multinaturalism*,[3] not those of other cultures (Descola, Viveiros De Castro, Ingold) or in technical-scientific practices (Latour) but in our modern day. In this way, multinaturalism shouldn't be viewed solely as the paradigmatic existence of multiple natures in different cultures (or "ontologies" as Descola referred to them) but also and most importantly as the syntagmatic presence, and the resulting contrast, of various ontologies within the same culture, which give rise to something akin to an *internaturality*, all of which reconsidered from a semiotic standpoint, in light of a conceptual apparatus, and a series of models for investigation, which reason not in terms of objective supposed realities but of plural significations. In other words, from a semiotic point of view, the so-called ontologies (however many they may be) are considered to be *effects of meaning* derived from precise *discursive dispositifs*. It is in the social discourse, the resulting dynamics of multiple subjacent textualities, that something like nature is produced, or rather, that an implicit cultural presupposition finds its *raison d'etre*, a presupposition that allows us to understand – and therefore define and at the same time profess – certain figures in the world as entities similar or dissimilar to us in terms of interiority and/or physicality. Animism, for example, is not an idea of nature in itself but one of the many possible ways in which the human subject, individual or collective, relates to psychic continuity and physical discontinuity with other living species. In oppositional terms, naturalism would be that silent presupposition that considers other living beings, animals or plants (even atmospheric phenomena) to be in psychic discontinuity and physical continuity with us human beings. It is in similar terms that we define totemism (physical and psychic continuity) and analogism (physical and psychic discontinuity). Starting from a certain number of differing textual realities, what interests us is the demonstration if and how the different discourses now manifest themselves, be they animist, naturalist, totemic, and so on, and how they intertwine in both a dynamic and efficient way. The media today, both old and new, are, from this point of view, an infinite treasure trove of textual manifestations of this kind of discourse: taking from the social culture to which they are accountable, produce discursive models that fall back onto that reality, modifying it to varying degrees, and more or less reconfirming it.

---

[2] In this regard, the work of Primo Levi is particularly relevant, in which the theme of "being human" is constituted in semantic opposition to being a "beast": cfr. Marrone (2016).

[3] For a broader view of this concept and its semiotic significance, see Marrone (2011), Marrone ed. (2012b), and Marrone et al. (2016). The reference here comes particularly from Latour (1999), Descola (2005), and Viveiros De Castro (2009). An excellent anthology of texts on this theme is Consigliere (2014).

When trying to operate within this immense field of research, the *point of view* of the animals, or, if you prefer, that of animality, proves immensely helpful. Why? Because such a theme allows us to face the issue of the relationship between nature and culture starting from a middle ground, from that existential condition that, by separating them, keeps them together; it is precisely not the animal but the animal culture.[4] The animal is, according to the naturalist perspective that is our own, everything that exists within nature, and it is the human being who must consider the problem of its own placement in these terms, carefully defining its own naturality and its own culturality. From an animist perspective (which is, as we will see, very often also our own), the problem is put forward in a very different way: between human beings and animals, there is no separation on the psycho-cognitive and emotional level; it is society that, pressing down the pedal of physical differences between humans and non-humans, tends to separate them, at times prioritising one, at times prioritising the other.

On one hand, as has often happened in Western thought, we find ourselves discussing variable thresholds, and the respective crossovers, present in the human/animal relationship.[5] On the other hand, as far as the semiotic paradigm is concerned, it is working towards an advance in research into this vast field of research known for some time now as "zoosemiotics". This term should not, however, be understood here in a tendentially naturalistic way, as a study of languages and the diverse forms of communication of various animal species, and their hypothetical "mind" or "conscience" as has been the case since at least Montaigne, through Sebeok, and continuing today.[6] However, as a wide-reaching semiotic analysis free from prejudicial distinctions, where it is not the essences ("animal", "human", etc.) that are pertinent, semiotically speaking, but the *reciprocal relationships* and *narrative transformations*, the human → animal transfigurations, and vice versa, animal → human. We must outline an investigation into animal discourse, social discourse *on* animals and, perhaps, *of* animals, just like those that for a while now have been referred to as the politics of animality or (and here comes our subtitle) animal cultures.

Two preliminary observations are necessary. First and foremost, it is important to remember how behind the age-old question of rationality and language in so far as a (possible) way of discriminating between human and animal, there is another conviction that is neither equal nor clear, which derives from an effect of connotation. Sometimes we have to understand how much of the animal is (also) rational and linguistic, with a mind or a conscience or emotion. At other times we must

[4] See the Treccani online dictionary, under "**animality**": "s. f. [from the latin. tardo *animalĭtas -atis*, der. di *anĭmal* «animale»]. The constitutive essence of the animal, the complex of qualities that belong to animal life (and in contrast to the spiritual tendencies belonging to humans) (D'Annunzio)".

[5] There is an enormous amount of debate around this theme. For a good overview, see Cimatti (2013).

[6] See the well-known synthesis by Sebeok Th (1968).

instead grasp the extent to which the human is bestial, where the bestial (as explained above) is instinct, and so violence, cruelty, and conflict are a given. It often happens that when searching for humanity, animality can lead us to forget that humanity is not intrinsically peaceful or reasonable but rather full of conflict, violent, and terribly cruel. From this comes a certain prejudicial, bleeding-heart attitude that obscures the (semiotically obvious) fact that human and social meaning is always created through conflict, struggle, and polemics. Secondly, when it comes to the so-called scientific approach to animality, it appears apparent that giving a place to an animal among the "kingdom" of nature (thinking it, defining it, etc.) means, at the same time, inserting it among the articulations of the social, the institutions of culture (kinships, affections and affiliations, strategies and conflicts, and so on). In other words, making it an object of knowledge is proposing a certain vision, and praxis, politics. In doing so, the current environmental, ecological, and animalistic demands, along with the ideologies of vegetarianism and veganism, have done nothing but reveal the naked king. There has always been a politics of animals: now it has its own eloquent and very determined spokespeople.

2 Faits Divers

We will start with a news article that is rather similar to the *faits divers* studied by Barthes (1963) and taken from the Italian newspaper, *la Repubblica,* on February 5, 2016. An article like many others found in old and new forms of media the world over.[7] This article was chosen because it offers a vision of nature – or rather, a conception of the relationship between humans and non-humans – that is anything but peaceful, highlighting with sufficient clarity the presence of an internaturality within the contemporary media discourse.

The article examines the case, considered exceptional by the newspaper reporting it, of a "blinding jealousy" between female orangutans. The journalistic value of the piece lies in the fact that such passion, considered by the newspaper to be typical of humans, can also be found among "anthropomorphic apes", thus bringing beasts closer to human beings, precisely because of this common "violent" and "criminal" behaviour. Here is the text:

> *The Case. The Orangutan and the Crime of Passion in the Forest.*
> *A scholar tells of the first execution of primates ever reported*
>
> Madrid. Crime of passion in the jungle. But human beings, for once, are not involved. Rather, they are the impotent and dismayed witnesses. If the orangutans are anthropomorphic apes, then we are now surprised to discover that they can be capable of violent and criminal behaviour just like people. The murder – the first case of which has ever been reported within a usually peaceful community such as that of these primates – happened in July 2014, but has only come to light now that the researcher from the University of Zurich,

[7] Some of these *faits divers* regarding the modern, problematic coexistence between humans and animals have been collected on the FB page: https://www.facebook.com/zoosemiotics/

> Anna Marzec, a direct witness of the unusual tragedy, has written about it in the journal, "Behavioral Ecology and Sociobiology".
>
> Tuanan Reserve, the jungle of Indonesian Borneo. According to the account given by *El País*, the brutal attack took place in a fit of rage, the victim a 35 year old adult female, Sidony. It was another younger female, Kondor, just 15, who started the attack. The male Ekko had been courting Kondor when Sidony passed by, followed by her son Sony, 4 years old. Ekko is distracted momentarily from his recent conquest and moves towards the older Orangutan, inspecting her sexually. Unconvinced, he retraces his steps and opts to start sexual relations with the young Kondor. However, seemingly blinded by jealousy despite the fact Sidony had, in the meantime, continued on her way, the young orangutan leaves her partner mid-coitus and attacks her. Punches, scratches, bites meted out with extraordinary violence, according to Anna Marzec, first by Kondor, who is immediately followed by Ekko who arrives to help her. "A coalition attack, prolonged and coordinated", with the male intervening only at intervals but inflicting the most serious wounds on Sidony's ravaged body. This continues for 33 long minutes, until the screams attract the attention of another male who manages to chase off the merciless aggressors. It is too late, however: Sidony died twelve days later from her wounds.
>
> A kind of crime that, even if it has precedents, has never been directly observed by the scientific community. A similar violence was discovered forty years ago among chimpanzees by Jane Goodall, who witnessed a feud between clans.

The "case", though rooted in scientific discourse (we'll come back to this), is sensational news. It is a classic "crime of passion" that takes place, however, "in the jungle", more precisely, in the sufficiently exotic reserve of Tuanan in Indonesian Borneo. It is newsworthy because there have never been similar cases of "execution among primates", an "unusual tragedy" that allows us to "be surprised to discover" that orangutans can "be capable of violent and criminal behaviour just like people". So, on the one hand, we have an event that "for once" does not involve human beings. On the other, the humans are hugely involved if only as a term of comparison regarding the behaviour of non-humans. If, as we know from Barthes (1963), every piece of news consists of a relationship being made between two informative elements, the enunciated event and the presupposed situation, the structure of this news is rather evident: to discover with surprise that the orangutans can also exhibit jealous and violent behaviour, "just like people", presupposes the fact that, on one hand, we assumed that they weren't and that, on the other, on an even deeper level, it was known that "people" have always been like it. Human beings are normally (naturally?) capable of similarly atrocious fits of blinding jealousy that lead to real, tragic executions (behaviour that can be described generously as bestial). Animals, however, we believed were not capable of this; we thought that they knew how to maintain a civil (that is to say human) level of behaviour. Has the world turned upside down? It might seem so.

In any case, this has all the hallmarks and more of the famous *aberrant causality* formulated by Barthes: not only is there among the motives of the jealous orangutan and the consequences of her violent act, a certain disproportion that renders the event abnormal, but, in addition, it all takes place in the world of the apes who, though anthropomorphic, were supposed to be free from such passions. As Barthes reminds us once again, proof of this lies in the shock (cfr. "surprised to discover") the event generates in the reader inscribed in the text (and among others), who cannot help but share that which, according to philosophers, is the primary reason for scientific knowledge in the world: the sense of wonder experienced (one would imagine) by the researcher from the University of Zurich, Anna Marzec, when witnessing this unusual tragedy, by readers of *Behavioral Ecology and Sociobiology*, and so forth through the chain of enunciations (something else we will come back to). The primary opposition is made between human beings who "for once weren't involved" and the orangutans, "anthropomorphic apes", protagonists of the sordid affair: where the former usually behave in a violent way because of blinding jealousy, the latter seem to be free from such characteristics. But such opposition is neutralised by the specific account of this particular crime of passion. Between humans and non-humans, there isn't all that much difference, given that even the latter are capable of the same horrors as humans. The animalist bleeding heart, posed because it is presupposed, takes a blow.

However, there may be another way of reading the article in question. Let's try to reconstruct the articulation of the enunciated tale. There are two antagonistic subjects, actualised as female orangutans: the first subject is Kondor, the assassin, an ape "barely 15 years old", target of amorous advances from Ekko, a male of the group, whose age is never mentioned (because, we suppose, it is considered of little importance compared to that of the females); its antisubject is Sidony, the victim, an adult ape some 35 years old, with her son in tow, Sony, just 4 years old. The two females are contesting an object of value, the orangutan Ekko, a false protagonist of the story because, as is clear, he is nothing more than an object in which a precise value is inscribed. It appears to be Ekko who acts: he courts Kondor and then deviates to Sidony, but returns to the first in order to copulate with her. His vague exploration of the world (which of the two females to choose?) reveals itself to be fatal, not for him, as he actually participates (though it is not clear why[8]) in the incredibly violent killing of Sidony, but for Sidony herself, victim of Ekko's amorous uncertainty and the extreme jealousy of Kondor. At a certain point, a second male orangutan intervenes, judging addresser with no name or age but to no effect: what is done is done. Sidony will die in a matter of days. So, despite the many tales ethologists, and the media with them, have told us about the terrible alpha males surrounded by crowds of females ready to copulate, in this strange group of apes, it is the females who are the real protagonists. Ekko, thanks to his indecision, perhaps his lack of experience, jumping from one female to another causes a terrible mess, before then helping Kondor kill the poor Sidony. The other male, though he manages to chase away the two furies, is wholly ineffective. The two women, however, are intensely active and efficient. Kondor reacts passionately and goes for the jugular, allowing herself to be carried by the fury with no moral qualms or doubt of any kind. Sidony, mature with her son, knows herself: she allows herself to be explored sexually by Ekko but doesn't seem particularly bothered. She has her young Sony to look after, she knows the world, she knows what importance to give to things in life, and she has family.

[8] This is a pertinent question for the ethologist.

Now, in their analysis of jealousy, Greimas and Fontanille (1991, pp. 167 onwards) begin to distinguish *rivalry* from *attachment*, demonstrating the strong internal dialectic. They observe that, if we look carefully, one cannot exist without the other, and they serve to strengthen one another. The former would occur in the jealous subject "if the event is caught before it can happen", if therefore "the rival's connection with the object has not taken place". The latter presents itself when "the event has taken place", and the jealous subject, "unless it attempts to take revenge", "turns to the object", despairingly, attaching himself to it in any way possible. In our case, things seem to be halfway between the two: "the event" in itself, that being Ekko's copulation with Sidony, has not, in a strict sense, happened, given that the inept male limited himself to "sexually exploring" the poor, ageing female. Instead the "event" happened with Kondor who, despite this fact, is the highly jealous subject: not only does she interrupt the sexual relations with Ekko, but, putting aside any possible future griping (attachment), she immediately moves into action, and – it must be said, in an entirely bestial fashion – she takes revenge with all her might, raining scratches and mortal bites down on poor Sidony.[9]

If Freud and Darwin's jealous apes, according to Despret (2002, chap. II), were uneducated and domineering animals that, however, achieved Victorian monogamy with their many females, according to the account of the Swiss researcher, the female orangutans in Borneo seem to have no ethical scruples. They mate, and heaven forbid they should be cheated on. And yet, this isn't exactly the case: this isn't a case of betrayal (the "event" didn't happen) or of "obscuring" (a typical passionate figure in the configuration of jealousy, according to Greimas and Fontanille). We could say, perhaps, that the "event" is, in this case, neither virtual (purely obscured by the jealous subject) nor realised (the "connection"), but perhaps we can describe it as *actual*, extremely possible but as something that has not yet happened. What blinds Kondor with jealousy, the newspaper surreptitiously tells us, is not Ekko's mating with Sidony, nor the vague possibility of this happening, but the fact that she had been ditched in the middle of a seduction that had already started, albeit for no more than a few minutes, and then being chosen only after the "sexual exploration" of her rival Sidony by Ekko, and all before her own eyes. This is where the wounded honour lies.

At this point we have to ask ourselves: what exactly is the value Sidony and Kondor inscribe in their object, in this case, Ekko? Evidently it isn't a sexual desire: Sidony has other things to think about; Kondor leaves mid-coitus. Ekko is not the chosen partner for a banal momentary coupling but something entirely different. Perhaps he represents a more abstract value, linked to a social identity that is almost entirely opposite to that of the two females involved. To possess him – or rather, to possess him *at this point in time* – confirms their own feminine way of being. Kondor hopes it enough that she is barely 15 years old, with all the youthfulness and

[9]The tale of the killer beating intensifies demonstrably when you mention the "33 interminable minutes" it lasted. If the exact calculation of time can't help but produce yet another effect of the real, the adjective "interminable" is problematic: interminable for whom? For Sidony? For her son? For Anna who is watching the scene?

naivety conceded her by her age, and she won't listen to reason: she is young, attractive, and that's it. No one should put in doubt the total romanticism of youth as an absolute talent; no one should think (not even for a moment) that another kind of sexual attraction could possibly exist in the world other than that of her and her marvellous age. Sidony has already moved past this phase, she can't fall for the first male who only has his youth to offer (Ekko, at the end, abandons her), but she does display a maturity that Kondor doesn't possess. She has a son, other responsibilities, another lifestyle. And this is what bothers Kondor to the point of blinding her: Ekko left her to go after Sidony, a sign that, somehow, the adult orangutan could have something more than her, and she, as if not good enough, is witness to the scene of sexual exploration – "the image is that which I am excluded from" was Barthes' (1977) response to seeing the beloved object making advances to a rival.

It could be said that we have exaggerated; we have excessively anthropomorphised the reasons behind the behaviour of these apes in Borneo, attributing a complex, if not contorted psyche, effectively, a pathemic logic. But the analysis didn't actually do anything more than incatalysing – making explicit – that which the text very clearly wanted to say. It is the text, in its powerfully semiotic complexity of an animist kind, that, in its telling of a story that is anything but simple, bestows the animal with a humanity, a psyche, a cognitive and passionate capability that is usually considered the sole virtue of human subjects. Furthermore, the violence towards a real antisubject, as opposed to a virtual or actualised one, is a very sophisticated cognitive operation, which brings into play elements and properties that far transcend the (bestial) desire for sexual coupling as an end in itself: youth is compared to maturity, the immediate sexual attractiveness to adult seduction, vivaciousness to motherhood, and thoughtlessness to the wisdom of a life that has already been lived fully. All of this leads to the construction of a rivalry that is not eminently sexual, nor even amorous perhaps, but all-encompassing and personal experience of the world and social identity. To reiterate, what Kondor finds unbearable, according to the newspaper looking for a scoop,[10] is that Sidony is a woman (pardon, an ape) who is fully realised, not simply a woman to have sex with, whilst she, however, is.

This is where the "surprise" experienced by the many enunciatees of this tale lies. The jealousy at play here is human because it is cognitive, intimate, and interior, and the beasts, adopting it, become humanised in the name of an animism assumed to be the complex outcome of a long narrative and passionate journey. Furthermore, these apes already had an age and their own names, and they lived in a reserve where nature, protected, is rebuilt each time: what else could we expect from them? Animalistic behaviour? Of course not.

We now need to consider the question of enunciation. As has been hinted at already, the article inscribes its own journalistic discourse under the auspices of a scientific one, positioning it as content of its own enunciation. The article is twofold: the "case" lies in the fact Kondor killed Sidony out of jealousy, but also in the fact that Anna Marzec, having witnessed the event, wrote about it in a scientific review, which was then cited in a famous Madrid-based newspaper, with reference

[10] In a time of "femicide", this is a juicy new version for journalists.

to a prominent authority on the matter, the primatologist Jane Goodall. So, between the enunciated content and the enunciatee, there is a long series of enunciators, or rather mediators-brands, whom the newspaper's enunciator directly quotes in order to acquire their veridictive authority:

- The scholar, Anna Marzec, direct witness to the scene between orangutans, researcher at Zurich ("according to Anna Marzec") [first enunciative brand].
- *Behavioral Ecology and Sociobiology*, which publishes her scientific article discussing jealousy among apes ("has written about it in the journal") [second enunciative brand].
- The Spanish newspaper, *El País*, transforms the content of the scientific article into general news ("According to the account given by *El País* ...") [third enunciative brand].
- The first enunciator from la Repubblica, which reports the news given in the Madrid-based newspaper, placing the account of the facts in Madrid itself [fourth enunciative brand].
- The final mention of Jane Goodall ("A similar violence was discovered forty years ago among chimpanzees by Jane Goodall, who witnessed a feud between clans."), an exceptional witness who "poetically" brings the text to a close in the same way it began [fifth enunciative brand].

On the one hand, in this way the enunciator produces a chain of internal references (from Goodall to *El País*, from *Behavioral Ecology and Sociobiology* to Marzec, to the University of Zurich) which ends up rendering the jungles of Borneo as an undeniable reality. On the other, it sets off the "ontological" presuppositions of its own discourse from all of these supposedly authoritative voices, making the enunciative origin of the epistemological "moral" of our story impossible to pinpoint. Who is it exactly that says a phrase such as "If the orangutans are anthropomorphic apes, are we now surprised to discover that they are capable of violent and criminal behaviour just like people"? It is impossible to establish with any certainty. What is important, however, is that it is believed.

3 Humanimals

The newspaper article we have chosen brings into play scientific and, in particular, ethological knowledge, albeit in an easily accessible way. This is an extremely interesting observation for us, as long as we do not take its contents as facts but treat them instead in exactly the same way as we did the journalistic content: as a point of reference constructed, internally or externally, by the same discourse. Of great help here is the work by Vinciane Despret (2002, 2004, 2007), who has published a number of important socio-anthropological analyses on ethological practices. Much like Latour and Woolgar (1979), who carried out ethnographic studies of scientific laboratories, developing an almost canonical praxis of the fabrication of facts (Latour 2005), Despret aims to do the same in the field of those studying animal behaviour. Understanding ethology as a scientific discourse rather than a natural science leads us to ascertain, as we said above, that the definition of the animal and its behaviour is not so much the result of observation and an epistemologically neutral reflection, but the assumption of an ideological and political position that is also ethical, aesthetic perhaps, and essentially semiotic. The scientific gesture is combined with real action and the values needed to produce it. So, for example, the Russian prince Pëtr Alekseevič Kropotkin, known theorist of anarchy, studied the animal world and so-called primitive populations to find a model for the construction of a future political utopia (Despret 2002).

According to Despret, and many other scholars working on the same themes, it is necessary to work on those "hybrid communities" that bring together humans and animals with variable relationships among themselves, starting from the most diverse circumstances and practices. In other words, the problem isn't studying animals but studying societies and "discovering" that they are not exclusively human, as we tend to think, but also animals, with the most diverse relationships with humans and among themselves. The ethno-ethological gaze thus leads us to clarify these complex entities that, alongside Donna Haraway, Despert calls "humanimals". In his introduction to a large conference on the subject, the scholar makes his position very clear:

> nous nous sommes préoccupé, non seulement des animaux avec lesquels des humains échangent des services, des informations et des affects ('espèces compagnes', animaux de service et animaux de ferme), mais aussi des animaux sauvages qui vivent libres et plus ou moins indépendants des établissements humains. Nous avons donc évoqué des animaux qui importent à certains entre nous, soit parce qu'ils partagent leur espace quotidien, soit parce qu'ils sont au travail avec eux, soit parce qu'ils les chassent, s'en défendent ou les protègent, soit enfin parce qu'ils en font un objet d'étude ou de réflexion. (Despret and Larrère 2014, p. 8)

This means, among other things, that the affirmation frequently used by philosophers – "we know nothing about animals" – only possibly has a meaning within a naturalist ontology that understands knowledge as the result of "pure" scientific experience which, in principle, gives rise to unique truths and indisputable laws. If instead, reinterpreting Despret, we look at it through a semiotic prism, this affirmation appears rather false. Firstly because, as we know from narrative grammar (Greimas and Courtès 1979), in the immanence of our concrete social practices, tacit and implicit forms of knowledge, and forms of valorisation, are implicated. Furthermore, such forms are, at any rate, made explicit by the analysis of the paths and values of narration. Since at least Propp, but perhaps even before Aristotle, we know that starting from the sequence of actions carried out by the actors involved – fictional or otherwise, it's not important – we can reconstruct the psychological motivations for the actions themselves, and therefore also the "interiority" of the characters, regardless of whether they are humans or non-humans. In other words, even if we know nothing about what the animals are thinking, or even if they think at all, we know full well what they do, with us and among themselves. This knowledge allows us to infer the "psychic" presuppositions of their actions, as well as the forms of sociality in which they operate. Kondor *docet*.

This form of knowing about the animal, according to Despret (2002), cannot be an "ascetic" knowledge, that of an external observer who fixes their own objectifying gaze on a supposedly universal and necessary nature, but the committed gesture of a subjectivity that intervenes in the field, bringing passions into play, alongside motive, has a worth beyond that of other methodologies. This is exactly where the famous experiences of Jane Goodall are situated, to whom Despret dedicates a great deal of time. Goodall refused to study the chimpanzees without having previously created some kind of emotional relationship with them. By infiltrating the social lives of these apes (their families, raising of young, etc.), she was able to overturn Darwin's (and Freud's) ideas on jealous apes, in which the two theorists maintained that one single male would dominate the entire group. Transgressing scientific protocols to reinvent a science filled with affection, a feminine science (according to Despret), Goodall led a sea change in primatology, but the chimpanzees, and she herself, also changed. Bringing into play the fact she was a sensitively involved woman rather than a coldly impartial scientist, Jane Goodall was able not only to understand the role of the female in the group and the centrality of motherhood in the same but her own maternal experience. She would often say that "the apes taught me how to be a mother". This is not just a line but a famous and extraordinary example of the world in which the relationship of knowledge, in the ethological discourse, is not between a subject and an object, but two subjects, and therefore also includes the intersubjective relationship between the two, in which each changes, or can change, in relation to the other. As a world shared between humans and non-humans in everyday and social life does exist, there is no reason why it should be placed within the parenthesis of scientific knowledge. The human gaze on the animal, strategic like every other, modifies the animal whilst at the same time modifying the human. By observing the animals, humans transform them, not just in their cognitive representation but in their ways of doing and being, in their values, their cultures of reference, and in doing so they transform themselves. In other words, transformations and the relative momentary identities are reciprocated according to the respective strategies and tactics. Think of parrots: according to Despret, it is not a case of stereotypically understanding if they know how to talk or not, but whether they have something to say, if they know how to do it, whether they have something interesting to say to one another. In this way, the problem shifts: the crux of the problem isn't, for ethologists, the faculty of language, but the set of tools for communication and values that are at play. Similarly, it seems they only began to understand the behaviour of crows when a research took them into his home, giving them an identity, a role in an interpersonal relationship with the researcher and his family.

4 Accelerated Evolution

When discussing primates and scientific knowledge, we must not forget Franz Kafka's academic ape. The reference comes from the famous short story entitled, *A Report to an Academy* (*Ein Bericht für eine Akademie*, 1917), in which an animal, more precisely, an ape, transforms into a man: a sort of accelerated parody of Darwin's theory of evolution. This story is the opposite of "Metamorphosis", in which a man becomes an animal. The phenomenon of *becoming animal*, as observed by Deleuze and Guattari (1975), is central to the work by the Bohemian author, a work in which the transformation itself and the modes of passage are much more important than the initial and final essences. Furthermore, in Kafka, the metamorphosis never fully takes place: there is always a leftover, a hybrid, a protagonist who is never fully human nor fully animal.

As Pezzini (1998) observes, in "Metamorphosis", when the story begins, Gregor Samsa already has the body of an animal. What happens over the course of the tale is the progressive acquisition of various animal "skills". He must learn to have the same passions, the same desires, the same tastes, the same plans of action, the same values as a cockroach. And when he has finally managed it, he allows himself to die. In "A Report", more or less the same thing happens, except that the ape, who is now almost human, starts to play along with the humans, despite no possibility of finding true freedom. Another pertinent difference between the two texts is that in "Metamorphosis" the discourse is in the third person, whilst in "A Report" it is written in the first: the ape is the one who tells his own story in the form of an academic report. But what is his story? It is the events of his capture in Africa, his transport to Europe on a boat in which Peter the Red (this is the name he gives himself) is constantly kept locked in a cage. In order to find a possible means of escape, a *line of flight*, he decides to slowly take on typical human traits, or rather, traits which he believes to be human ways of living, starting with his constitutively wild-eyed, bewildered look, which is very much other. The ape progressively understands (or believes that he has understood) what men love to do: spit, smoke the pipe, and drink lots of alcohol. So, in order to be accepted by men, and – he hopes – be freed, the ape apes his captors; he does exactly what we stereotypically expect apes to do: imitate the behaviour of those around them. So he starts spitting, smoking a pipe, and, not without difficulty, swigging back alcohol. At work here is a double regime of veridiction: the ape believes his actions are typical of men, he is trying to become human, and he does so to be accepted by the humans keeping him prisoner because he is an ape. But the men observe him, they find him incredibly amusing, believing that he is doing things typical of an ape, and so they continue to keep him in a cage. From which – as with Jane Goodall and her chimpanzees – comes the double metamorphosis: Peter becomes a showman, an actor, whilst the men who watch him are no longer unscrupulous hunters but almost kind spectators, people who allow themselves to be enchanted by a caged ape. The prey has had to – and known how to – transform itself, not into a generic man but into a comedian in order to transform his predators into spectators. But at the same time, the predators have transformed the prey: from a wild ape, Peter has become other, he has gone up a level, he has evolved, specialising in something useful, something lucrative, and they benefit from it. So, when Peter the Red finally arrives in Europe, he is sent to a circus, where he continues to work in order to "train himself", thus becoming more human (or an ape that imitates humans more effectively). And he manages it: Peter becomes a celebrity (remember that he is the only narrator) and takes private lessons to become an exemplary man (or an exemplary circus ape). Finally, at the height of his career, he was invited to a university to give an academic report on his own curious story of evolution through adapting to his environment. As with Despret's ethno-ethology, the transformation is not unilateral but reciprocal, and the animal's domestication is not a loss of his own natural state but a further naturalisation that takes place in tandem with the transformation of the human beings.

The most significant aspect comes nonetheless from the fact that the whole story is read by an enunciatee who follows it step by step, phenomenologically recognising the steps of the ape's discourse. And he starts doing so, obviously, from the title. The reader must therefore progressively understand that it is not, perhaps, a real academic report given by a professor, but the academic discourse of an ape who has, extraordinarily, rapidly evolved, and it is thus the learned testimony of a subject who demonstrates, whilst ridiculing, Darwin's theory of evolution. But are we really dealing with a scholar, a professor, given that it is the outcome of an evolution from an ape-like state? Or is this an ape perhaps retelling his real-life experiences after having become an actor? Or, perhaps it is neither a professor or an ape nor an academic report, but a circus act in which an ape, or rather a man aping an ape, tells a fictional comic tale of an ape who takes Darwin's theories a little too seriously and now believes himself an ape who has transformed into a man. Installed here is a regime of belief that Todorov (1970) would have called *fantastical*: that of an a priori unsayability and therefore of a knowledge that adapts to different contexts, that objectively refutes the asceticism of the naturalist (in both senses of the word) in order to live with the entities – the humanimals – that he is studying. Who is talking in the text? Who are they talking to? Any answer can make sense; none of them are certain.

Finally, it should be observed that what happens in this tale is the exact opposite of the experiences of Gregor Samsa: the ape becomes a man, thinks like a man, has almost taken on the men's moral code, but he still has the body and affectivity of an ape. He is hairy and his passions are not very human at all: they are bestial. He doesn't, for example, feel any shame in pulling down his trousers to show his scars. And when he wants to have sex, he has no doubts: he turns to the female apes, with a great desire to placate, but with a certain *prudery*:

> In the evening I almost always have a performance, and I could hardly be more successful. When I come home late at night from banquets, from scientific societies, or from social gatherings in someone's home, a small half-trained female chimpanzee is waiting for me, and I take my pleasure with her the way apes do. During the day I don't want to see her, for she has in her gaze the madness of a bewildered trained animal. I'm the only one who recognizes that, and I cannot bear it.

References

Barthes, R. (1963). Structure du faits divers. In *Id Essais critiques*. Paris: Seuil.
Barthes, R. (1977). *Fragments d'un discours amoureux*. Paris: Seuil.
Cimatti, F. (2013). *Filosofia dell'animalità*. Rome-Bari: Laterza.
Consigliere, S. (Ed.). (2014). *Mondi multipli*. Roma: Youcanprint.
Deleuze, G., & Guattari, F. (1975). *Kafka, pour une littérature mineure*. Paris: Minuit.
Descola, P. (2005). *Par-delà nature et culture*. Paris: Gallimard.
Despret, V. (2002). *Quand le loup habitera avec l'agneau*. Les empêcheurs de penser en rond. Paris.
Despret, V. (2004). *Hans, le cheval qui savait compter*. Les empêcheurs de penser en rond. Paris.
Despret, V. (2007). *Bêtes et hommes*. Paris: Gallimard.
Despret, V., & Larrère, R. (Eds.). (2014). *Les animaux: deux ou trois choses que nous savons d'eux*. Paris: Hermann.
Greimas, A. J., & Courtes, J. (1979). *Sémiotique. Dictionnaire raisonné de la théorie du langage*. Paris: Hachette.
Greimas, A. J., & Fontanille, J. (1991). *Sémiotique des passions*. Paris: Seuil.
Latour, B. (1999). *Politiques de la Nature*. Paris: La Découverte.
Latour, B. (2005). *La Science en action*. Paris: La Découverte.
Latour, B., & Woolgar, S. (1979). *Laboratory life: The construction of scientific facts*. London: Sage.
Marrone, G. (2011). *Addio alla Natura*. Torino: Einaudi.
Marrone, G. (2012a). *Stupidità*. Milano: Bompiani.
Marrone, G. (Ed.). (2012b). *Semiotica della natura (natura della semiotica)*. Milano: Mimesis.
Marrone, G. (2016). *Levi analogista*. Riga, forthcoming.
Marrone, G., Ferraro, G., Giannitrapani, A., & Traini, S. (Eds.). (2016). *Dire la Natura*. Roma: Aracne.
Pezzini, I. (1998). *Le passioni del lettore*. Milano: Bompiani.
Sebeok Th, A. (Ed.). (1968). *Zoosemiotics*. Bloomington: Indiana University Press.
Todorov, T. (1970). *Introduction à la littérature fantastique*. Paris: Seuil.
Viveiros de Castro, E. (2009). *Métaphysiques cannibales*. Paris: Presses Universitaires de France.

On the Logic of Animal Umwelten: The Animal Subjective Present and Zoosemiotics of Choice and Learning

Kalevi Kull

Abstract The aim of this study is to demonstrate the mutual relationship between the classes of signs, mechanisms of learning, and types of umwelten. This framework is necessary in order to describe the animal ways of meaning-making in the context of various forms of semiosis. We state that semiosis only occurs in the internal present. An account of the linkage of vegetative, animal, social, and cultural umwelten with corresponding classes of signs and mechanisms of learning (imprinting, conditioning, imitating, conventioning) is provided. We also formulate two rules of semiotic development that characterise the ontogeny of the umwelt.

1 Understanding Animals

Understanding animals – as well as any other organisms – means being able to describe their world, to know what they can know. We are going to describe in this essay some steps of analysis of animal umwelten.

The semiotic approach to the analysis of animal behaviour (e.g. Sebeok 1968, 1972, 1977, 1988, 1990; Marler 1978; Maran et al. 2011; Martinelli 2010; Kull 2014; Brentari 2015; Kull and Torop 2011) is not limited to the description of external forms of communication and behaviour but provides scientific tools for the study of internal aspects of behaviour – of the animal umwelt.

We develop here three statements:

1. Umwelt is related to the internal present; semiosis takes place in the Now.
2. The major types of umwelt depend on the major mechanisms of learning which are related to the major classes of signs.
3. More complex types of signs require a "longer" internal present.

This leads to the formulation of the rules of umwelt development.

K. Kull (✉)
Department of Semiotics, University of Tartu, Tartu, Estonia
e-mail: kalevi.kull@ut.ee

G. Marrone, D. Mangano (eds.), *Semiotics of Animals in Culture*,
Biosemiotics 17, https://doi.org/10.1007/978-3-319-72992-3_10

2 On Our Method

How can we acquire knowledge about the world of another being? It occurs that this is possible by observations and experiments, in connection with semiotic explanatory modelling (where the latter is crucially important, if not obligatory). How does this work?

1. *Precision or the faculty of discrimination.* There exist experimental methods to discriminate between the different distinctions animals are capable of making or to learn to make. The whole school of this kind of approach was established by Jakob von Uexküll and his collaborators in the *Institut für Umweltforschung* in the late 1920s and 1930s and followed by many studies in experimental ethology and zoopsychology. Here belong the experiments about the capabilities of the sensory-motor apparatus and the perception-action cycle of organisms – their capacities to discriminate between smells, colours, sounds, patterns, angles of vision, etc., as related to their regulation of movements or other kinds of actions. These also include Uexküll's studies on minimal size of space (*Orten*) and time (*Momenten*) in different species (Uexküll 1928; Magnus 2011). Thus, we can study the differences that make a difference for other beings. However, this is certainly not enough in order to describe their umwelt.
2. *The logical structure or faculty of interpretation.* The key to describing an umwelt as a whole is the establishment of logical (semiotic) operations or types of interpretation the organism under study possesses. These operations, or types of semiosis, or types of interpretation are seen as the organisers of the animal umwelt, as well as the operators or constructors that are building the model of the world for the organism. Moreover, the same operators can be used also by researchers as the model builders. Without having a direct access to the umwelt of another species, we are nevertheless capable of deducing its features on the basis of their organising principles and the logical operations that species possesses.

When using the term "logic" here, we should stress that this is logic in a considerably extended sense. This is logic that may not require the logical operations as described in formal logic. It can be considerably simpler than that; it is mostly preformal logic. The one who generalised the concept of logic as including all possible forms and phenomena of meaning-making was Charles S. Peirce. Thus, we are rather asking how the first forms of logic used by living beings emerge from the world of non-logic. As a research programme, it is close to the idea of the naturalisation of logic (as described, for instance, by Lorenzo Magnani 2015).

Hence, without aiming to precisely feel *how it feels* to be another being (as was famously asked by Thomas Nagel 1974), we are able to *comparatively describe several features* of *how they feel*.

While the Saussurean approach has led to studying codes, the Peircean approach tends to lead to study the processes of learning (or habit taking). Both of these phenomena can be studied using various methods. As Jesper Hoffmeyer (2003, p. 2663) has stated, "thus, biosemiotics does not exclude the traditional methods of biological investigations. Instead, it interprets the meaning of results differently and suggests new experimental settings according to the altered theoretical understanding". This means that experimental semiotics, including experimental biosemiotics, is possible (see, e.g., Kashima and Haslam 2007; Galantucci and Garrod 2012). However, we still need to be very careful in drawing conclusions about meaning-making from the data of observed behaviours, because the sign relations of others are not directly observable; what we study is largely the internal life of organisms.

3 Semiosis and Its Implications: The Subjective Present

Semiosis, in other terms, is interpretation – including all its forms: perception as interpretation, action as interpretation, translation as interpretation, signification as interpretation, and meaningful communication as interpretation.

Interpretation, in all these cases, has some obligatory features. The main one of these is that it assumes the existence of options. This is because interpretation as a semiotic process differs from physical (non-interpretative) interaction by the choices it can make. Interpretation assumes a certain arbitrariness, a contingency, since it only works if there are options, which means if there exists some freedom. Neither determinism nor a fixed probability distribution provides the freedom necessary for enabling interpretation.[1]

Options can only be plural and simultaneous. A single possibility is not an option, by definition. For the same reason, the options cannot be just sequential – then they would be single, at each moment.

This reasoning necessarily implies that interpretation requires a simultaneity of options. Since options in order to be options have to be recognised, and since recognition cannot be momentary, it implies that the simultaneity has to have duration. This is the feature of what William James (1893) has called the *specious present*, or subjective now, and what we here call the internal present.[2] Moreover, "it was Karl Ernst von Baer (1864) who came up with the concept of 'moment', which is supposed to be the longest time interval to be objectively measured without apparent duration. Baer suggested that different organisms presumably have different moments if measured by external means" (Pöppel and Bao 2014, p. 244).

An important consequence we arrive at here is that the existence of *the specious present is coextensive with semiosis*. This means that semiosis is what happens in the present and only in the present. In other terms, representamen, object, and interpretant occur simultaneously. Semiosis takes place in the Now.[3]

[1] See also Kull (2015).

[2] Until no additional distinctions are made, we use the terms specious present, internal now, and subjective present, or just Now, as synonyms.

[3] We leave here aside the discussion about whether semiosis is a product of nowness or the Now is a product of semiosis. On the one hand, we could hypothesise that semiosis is responsible for

Interpretation in the situation of presented options always includes an aspect of novelty and unpredictability. This is because the situation of true choice as a part of interpretation assumes the unavailability or inaccessibility of algorithm that would apply for the given situation. "The present in our common experience always contains a part of radical novelty, a genuinely new part", observes Jacques Coursil (2015, p. 232). This is evidently a universal characteristic of Now in all living beings.

The atemporal aspect of semiosis has not received much attention in semiotic studies.[4] However, as Thomas J. Bruneau (1996, p. 104) says, "the constitution of the semiotic 'now' is what gives nonverbal activity its potential communicative or signal quality".

Another scholar who has approached the relationship between semiosis and time and wrote a series of works on the phenomenon of time was Julius Thomas Fraser (1978, 1999, 2001). He writes (1999, p. 64, 65), "The life process gives rise to a type of continuity or permanence that is qualitatively different from what one finds in the physical world. It is called intentionality. [...] The creation and maintenance of the organic present is a necessary and sufficient condition of life. [...] Readiness, this chronic incompleteness of life, is the primeval form of *need*". And in Fraser (2001, p. 269): "The physical universe is nowless".

The internal now has received considerable attention in experimental psychology (Pöppel 1997; Pöppel and Bao 2014; see also Varela 1999). However, there are not many comparative studies about the animal internal present yet that would follow the early ideas of Baer (1864) and Uexküll (1928).[5]

4 Semiotic Mechanisms of Learning

Interpretation, if it modifies memory, is obviously a form of learning.[6] In other words, semiosis, if it leaves a trace, is learning. This relationship between semiosis and learning provides a key to approach the study (and the description) of semiosis from studying the processes of learning.

There do not exist many results yet that relate the semiotic classes of signs with corresponding physiological mechanisms. The problem itself has been formulated,

---

nowness as such. On the other hand, the expansion or shrinking of the Now may influence the available complexity of sign processes for the whole organism. Some evidence for the latter comes from the observations of amnesia.

[4] Although Ferdinand de Saussure's concept of synchronicity can be seen as analogical to this. See also Bruneau (2007).

[5] About Uexküll's view on an organism's time, see Magnus (2011).

[6] We could also define memory as acquired scaffolding and learning as shifting the scaffolding in result of interpretation. Scaffolding in this sense may consist of codes.

however, more than once in neurosemiotics (Müller and Wolff 2003), in psychosemiotics (Michon et al. 2003), in cognitive semiotics, and of course in biosemiotics.

Knowledge is acquired by learning.[7] Learning, from a fundamental semiotic point of view, is the creating and establishing of new sign relations or modifying the existing ones. Different types of sign relations can be products of different mechanisms of learning. We review below the main types of learning mechanisms and describe their relationship to the different layers and to the whole structure of the umwelt.

Learning, as defined above, can obviously be of varying complexity. Several types of learning have been described. For instance, Bruce Moore in his review (Moore 2004) offers an evolutionary cladogram linking nearly 100 forms of learning, including no less than 5 imprinting processes, 11 varieties of Pavlovian conditioning, 10 of instrumental conditioning, and 8 forms of mimicry and imitation.

We use in the following a general approach to sign typology, which assumes that more complex signs may include simpler ones, and each of these is based on a certain physiological mechanism. This recalls the approach used by Terrence Deacon (1997; 2011).

Thus, we propose here a correspondence between the mechanisms of learning and types of signs. Since our suggestion is preliminary and rather hypothetical, we use here a classification that includes only four types of learning and is accordingly limited to four classes of signs.

4.1 *Imprinting and Icon*

The simplest form of learning would be the one that just makes a sign connection with something *one*. Among the options, one is chosen. We use for this type the name *imprinting*. In case of imprinting, the temporal coincidence of anticipation and option is required.

Following the Peircean tradition, the simplest sign relation can be called iconic. An iconic sign is then a sign which refers to the existence of something. Iconic semiosis is the process of recognition, and it can be described as a reference to something that fits into the recognition window.[8]

Iconic relations imply classification. The recognition process constructs similarity as based on indistinguishability in the context of the recognition window. Thus, that is not the mind-independent similarity, but the similarity as mediated and constructed via the operation of the recognition window, i.e. the mediation-created similarity.[9]

---

[7] On the biosemiotically extended concept of knowledge, see Vehkavaara (1998).

[8] On the concept of the recognition window, see Kull (2016).

[9] Cf. Eco's (1999) account of primary iconicity.

4.2 Conditioning and Index

In the case of conditioning as a form of learning, the coexistence of *two* icons provides a condition for creating a link – a new relation – between them. The icons have to be temporally close enough, in order for the integration to be possible. The link itself may become stronger as a result of repetition. This means that the link – the index – represents correlation between icons.

An index is a relation between two signs – a sign which refers to another sign: i.e. a sign which refers to correlation (correlatedness) between icons. An index certainly does not require a causal relation between the objects (as is sometimes stated) – correlation is enough.

There exist many forms of conditioning. These include classical (Pavlovian) conditioning, operant conditioning, etc. However, all these can be seen as constructing indexes.

In order to fix a link, a mediator is required. Here the nervous tissue provides a powerful means for finding correlations and establishing connections between very different and otherwise qualitatively or spatially separated activities. This is particularly so because a nerve cell may have around thousands of neighbours, while any parenchymal cell has them a hundred times less, which makes the possibility for selecting and establishing new connections considerably easier in the case of neural tissue. Accordingly, conditioning is commonly related to the organisms who have a nervous system – i.e. animals.

4.3 Imitating and Emon

In the case of imitation as a form of learning, the link or sign relation that should be made is more complex than in the case of conditioning. It requires a connection between a certain behaviour perceived in two ways and the same behaviour created. Imitation requires the linking of proprioceptive and exteroceptive recognition of a particular action. On the one hand, due to proprioception, or feedback from the organs of action, a link between perception of behaviour and the action that produces this behaviour in the same organism is commonly established by the functional cycle. Via conditioning, the proprioceptive recognition with exteroceptive recognition, a link between, e.g., seeing one's behaviour and making (producing) it can be developed. For instance – an animal can see a movement of what is recognised as a leg. Without yet distinguishing between one's own and another's leg while seeing it, there is thus a link between seeing a movement of another's leg and operating one's own leg – thus, imitation can happen.

The sign relation produced by imitation is more complex than the index and less complex than the symbol. We term this type of sign as *emon*. An emon is a sign which refers to the identity between signs.

The distinction between icons and emons and introducing the row *icon-index-emon-symbol* solve the paradoxical usage of the sequence index-icon-symbol (instead of the original icon-index-symbol) that can be found in some semiotic theories.[10]

The word *emon* as proposed here for the name of the sign type can be seen as a derivation from the Greek ημων (ἡμῶν), meaning "company". Occasionally, the term *emon* has been used by Paul Nemirovsky to denote an element of emotional or aesthetic information (Nemirovsky 1999; Nemirovsky and Davenport 2002).

The emon as based on the capacity of imitation may co-occur with the existence of emotions and the phenomena of empathy. There is evidence for the existence of capacity for imitation in several species of mammals and birds, while there is almost none in non-vertebrates. The ability for imitation is what is necessary for social learning (e.g. learning certain behaviours from parents or other specimens of the community). Accordingly the availability of emonic signs may be limited to vertebrate animals.

4.4 Conventioning and Symbol

Establishing a connection on the basis of convention (e.g. naming) can also be a version of learning – though a more complex type than the earlier three. It assumes a capacity for freely combining and recombining different signs.

The symbol is a relation that is based on the capacity of intentional (re)combination. This enables the creation of connections that may not depend on any specific features the objects combined possess. Hence the symbolic relation can be detached from similarities or correlations or causal relations the objects may have.

Thus, one can say that the symbol is the sign that refers to a difference between signs. This allows the interpreter to learn that particular sign is the *sign* of something else, i.e. to have a concept of the sign as such (see Deely 2005; Rattasepp and Kull 2016).

These features of symbols are precisely those that enable an agent to freely create sentences as combinations of signs, and produce language, as it is common in humans.

---

[10]The emon as a sign based on imitation may seem to have some similarity with the Richard Dawkins' concept of meme. However, there are several fundamental reasons why these are deeply different. The most important difference concerns the basic logical incongruence – the meme is defined as being based on replication (copying), while the emon is defined as being based on the interpretation process. From this, many consequences follow. The modifications a meme can have are based on random mistakes in copying; the agent has no role in meme theory. As different from this, the interpretation process as the one on which imitation is based is dependent on the choices agents make. Another aspect is that many writings in memetics mix up the symbolic and presymbolic levels of signs, even if they mean that memes are signs (what they often do). Thus, memetics is just a not-yet semiotics.

5 The Types of Umwelten

The availability of an additional mechanism of learning can be responsible for the next level in the modelling and construction of the umwelt. The jumps to the new level of semiotic capacities have been characterised as semiotic thresholds (Eco 1976; Deacon 1997; Rodríguez et al. 2017). On this basis, the major types of umwelten can be described (see also Kull 2010).

An important precaution, however, is appropriate. This concerns the implications made on the basis of solely external observations. Externally, very similar behaviours can be produced by fundamentally different internal mechanisms. For instance, this concerns the external similarity in social (including eusocial) behaviour between some vertebrate and invertebrate animals. Moreover, this includes many examples of similarity between the behaviour of systems with artificial intelligence and living beings. Navigation of a self-driving car and a car with a human driver can externally be similar even while the way this is achieved is not at all alike. So far, computers do not possess a subjective Now (or, e.g., emotional umwelten) regardless of the ways they behave.

Another precaution concerns the typology of umwelten we provide. It is evident that there are more types of learning and accordingly more types of umwelten. Therefore, what we describe is just a rough account of some forms, with an aim to demonstrate at least the possibility of studying the biology of knowing,[11] together with making the next steps in this research, after Uexküll, Maturana, Piaget, Hoffmeyer, Deacon, and others.

5.1 *Vegetative Umwelten*

The simplest imprints are evidently single qualities. We can identify these as qualia or qualisigns. An umwelt that does not include anything else than qualisigns, an umwelt that consists merely of separate acts of recognition-action, cannot represent a direction or movement, since there is nothing that would connect the qualisigns to a compound image. If there is no sign-to-sign connectedness due to the absence of associative learning, then this is an umwelt in which space cannot be constructed.

We call such umwelt *vegetative*. This is because the type of learning that creates it may not require even a nervous system.

A vegetative umwelt is an umwelt without subjective space or subjective time (however there can temporarily be a subjective Now). Being as simple as this, it nevertheless can be responsible for collective phenomena like the organisation of swarms or structures based on family resemblance (see Kull 2016).

[11] A main question semiotic biology is asking can be formulated as "what do the organisms know"? Cf. Kull (2014), Burghardt (1997, 2005), Bateson and Laland (2013), Hoffmeyer (2008), Esbjörn-Hargens (2017), and Mendelson et al. (2016).

5.2 Animal (and Spatial) Umwelten

The availability of indexes in addition to icons (as a result of associative learning) makes a huge difference in the representation of the world.

A particularly impressive faculty concerns the construction of space. At first, it may not resemble much of the spatial cognition we humans are used to. The construction of minimal space may start from connecting spots either in rows or networks, but already this, as linked to the acts of moving the body, allows the organism to search the way back to the place already visited, on the condition of course that the difference between the spots as perceived is detailed enough in comparison to the heterogeneity of the habitat.

Due to the capacity for associative learning, it makes sense to develop multicellular perceptual organs and to make use of sensory and motor categorisations. The functional cycles supplied to associations can enable an agent to establish relations of angle and distance and thus to build cognitive maps.

Thus, an animal umwelt provides a possibility for a moving organism to stepwise find the way back, to return to the place where it has already been, as different from a vegetative umwelt that may not provide this capacity. Jakob von Uexküll has described this as a phenomenon of a "familiar path" (Uexküll 2010 [1934], p. 98). Consequently, for animals (as different from mobile protists or bacteria) it makes sense to build a nest, as for instance several insects and spiders do.

5.3 Social (and Emotional) Umwelten

The operation of imitating provides a powerful mechanism of amplification. This can work at various levels of behaviour. For instance, reciprocal imitation enables the emergence of play. The capacity for imitation is also the basis for social learning and accordingly for social inheritance. As a consequence, group differences in behavioural rules can appear that may remind one the cultural or subcultural differences in many ways.

Another significant feature that seems to be connected to this level of umwelt is the existence of emotions. Emotions provide the level of categorisation that delimits or channels the choices in establishing new relations simultaneously for all modes. Emotions as linking systems also organise the whole syndrome of behaviour. Moreover, together with emotions appears the faculty of empathy that considerably enhances the ways of cooperation. As said before, these phenomena are rather widespread among vertebrate animals.

5.4 *Cultural (and Lingual and Narrative) Umwelten*

The neotenic change that was accompanied by overcoming the symbolic threshold in early humans turned in the organisation of the umwelt via the means of language. The symbolic semiosis itself is evidently responsible for the capacity of counting and (re)ordering (manipulating) various things following a given example.

A crucial change accompanied by symbolic reference is the origin of the operation of negation which can exclude the third. The oppositions available at the pre-symbolic levels do not exclude the third. The introduction of negation enables the development of syllogistic logic in lingual animals.

A particularly important consequence of symbolic reference is the combinative reordering of images which results in the constructive representations of past and future. Together with the construction of narratives (earlier levels of semiosis being incapable of this), it creates the possibility for imaginable travelling in time or chronesthesia (this concept has been introduced by Endel Tulving – see Tulving 2002). Accordingly, a true semiotics of time begins here, with symbolic semiosis.

Time travel (or daydreaming) masks much of the external perception otherwise available in the animal umwelt. This often results in a certain decrease in attentiveness towards the current surrounding, causing to some extent alienation from life in the present.

The powerful structuring of the umwelt with the help of language diversifies the manipulative possibilities. It creates many conceptual differences, which tend to lead to the enrichment of the emotional world.

The capacity to construct, imagine, and represent the other (the unavailable, the absent) and compare it with the current can also considerably enrich the possibilities and modes for suffering. As observed by Temple Grandin (Grandin and Johnson 2005), humans may have much more varieties of pain than animals. This means that possessing a linguistic faculty opens up the richness of suffering for an organism.

A distinct feature that the language faculty gives origin for is the capacity for true violence (see Kull 2011). Will and the ability for mindreading give humans the capacity to make intentionally bad decisions that may hurt someone. Thus, true violence – torture – is a result of language. This is radically different from animal predator-prey relationships which cannot be considered as violent because they cannot include an intention to torture (Weber 2016). However, it is not only the world of suffering that enlarges – all the emotional world, including the modes of joy, becomes more elaborated and more diverse (see also Deacon 2006).

6 The Rules of Semiotic Development

The first, a rather obvious rule[12] of semiotic development states that in the ontogeny, the change in complexity of meaning-making is related to the availability of types of semiosis. The growth in complexity of the umwelt is thus parallel to and a result of the development from iconic to indexical to emonic to symbolic semiosis. As a result, an umwelt can change during ontogeny from nonspatial to spatial to temporal.

Semiosis occurs in the Now. The second semiotic rule states that the movement from iconic to indexical to emonic to symbolic semiosis is taking place in connection with the widening of the internal present (the specious present). The advancement in the complexity of semiosis is dependent upon the extent of subjective present. Or, in other words, intellectual advancement is related to the expansion of the Now.

7 Concluding Remarks

Semiotics is a generalisation of linguistics, of logic, and of phenomenology – it also includes nonlinguistic sign systems, prelinguistic interpretation, and nonconscious phenomenology. Therefore, the developments in the areas that study non-human sign processes – zoosemiotics and the entirety of biosemiotics – are crucial for semiotics as a whole. Not only zoosemiotics, but the whole of semiotics, is coming to the next level, after its restructuration since 1990s. This is due to developments after post-structuralism and to recent results in Peircean studies and in biosemiotics, in particular.

What would deserve more attention in semiotics is the distinction between new and repeated sign relations, the (subjective) atemporality of semiosis, and the non-algorithmic nature of interpretation. Naturalisation of these features is a challenge for contemporary semiotics. Accordingly, an empirical grounding of semiotic theories (evidence-based and experimental semiotics) is important to discuss, particularly in the area of prelinguistic sign systems, as studied by zoosemiotics (which is a part of biosemiotics).

Sign relations are acquired or learnt relations; therefore, models of semiosis should be related to models of learning. We could assume that major sign types are produced by different types of learning mechanisms, and here we suggested a possible model for this.

---

[12] By using the term "rule" instead of "law" here, we want to emphasise that semiotic regularities as the ones produced by living beings (as nonuniversal regularities, i.e. prone to exceptions) are fundamentally (and ontologically) different from physical laws that are universal and not created by anybody.

Computational models of learning do not usually pay attention to meaning-making. The latter, however, can be possible via semiotic modelling that describes the situation of confusion and incompatibility as necessary in semiotic learning.

An empirical taxonomy of mechanisms of learning tends to deviate from Peircean classification. Building such a taxonomy may use a generalisation of the Lotmanian insight about non-translatability as a component of meaning-making and learning.

As we hope, the formulation of the rules of semiotic development will attract more research to study the ontogeny of semiosis. Pointing out the importance of the subjective present for any semiosis, we expect to get closer to the characterisation of life itself.

References

Baer, K. E. von (1864). Welche Auffassung der lebenden Natur ist die richtige? Und wie ist diese Auff assung auf die Entomologie anzuwenden? In K. E. von Baer (Ed.), *Reden gehalten in wissenschaftlichen Versammlungen und kleinere Aufsätze vermischten Inhalts* (Vol. 1, pp. 237–284). St. Petersburg: H. Schmitzdorff.

Bateson, P., & Laland, K. N. (2013). Tinbergen's four questions: An appreciation and an update. *Trends in Ecology and Evolution, 28*(12), 712–718.

Brentari, C. (2015). *Jakob von Uexküll: The discovery of the umwelt between biosemiotics and theoretical biology* (Biosemiotics 9). Berlin: Springer.

Bruneau, T. J. (1996). Subjective time, social interaction, and personal identity. In H. B. Mokros (Ed.), *Interaction and identity. Vol. 5: Information and behaviour* (pp. 97–115). New Brunswick: Transaction Publishers.

Bruneau, T. J. (2007). Time, change, and sociocultural communication: A chronemic perspective. *Sign Systems Studies, 35*(1/2), 89–117.

Burghardt, G. M. (1997). Amending Tinbergen: A fifth aim for ethology. In R. W. Mitchell, N. S. Thompson, & H. L. Miles (Eds.), *Anthropomorphism, anecdotes, and animals* (pp. 254–276). Albany: State University of New York Press.

Burghardt, G. M. (2005). *The genesis of animal play: Testing the limits*. Cambridge, MA: MIT Press.

Coursil, J. (2015). Hidden principles of improvisation. *Sign Systems Studies, 43*(2/3), 226–234.

Deacon, T. (1997). *The symbolic species: The co-evolution of language and the human brain*. London: Penguin.

Deacon, T. (2006). The aesthetic faculty. In M. Turner (Ed.), *The artful mind: Cognitive science and the riddle of human creativity* (pp. 21–53). Oxford: Oxford University Press.

Deacon, T. (2011). The symbol concept. In K. R. Gibson & M. Tallerman (Eds.), *The Oxford handbook of language evolution* (pp. 393–405). Oxford: Oxford University Press.

Deely, J. (2005). The semiotic animal. In R. Williamson, L. Sbrocchi, & J. Deely (Eds.), *Semiotics 2003: "Semiotics and National Identity"* (pp. 111–126). New York: Legas.

Eco, U. (1976). *A theory of semiotics*. Bloomington: Indiana University Press.

Eco, U. (1999). Kant and the platypus: essays on language and cognition (A. McEwen, Trans.). In *A harvest book*. San Diego: Harcourt

Esbjörn-Hargens, S. (2017). Animal worlds: The importance of biosemiotics for integral ecology. In S. Mickey, S. Kelly, & A. Robbert (Eds.), *The variety of integral ecologies: Nature, culture, and knowledge in the planetary era* (pp. 279–302). Albany: State University of New York Press.

Fraser, J. T. (1978). *Time as conflict: A scientific and humanistic study*. Basel: Birkhäuser Verlag.

Fraser, J. T. (1999). *Time, conflict, and human values*. Urbana: University of Illinois Press.
Fraser, J. T. (2001). The extended umwelt principle: Uexküll and the nature of time. *Semiotica, 134*(1/4), 263–273.
Galantucci, B., & Garrod, S. (Eds.). (2012). *Experimental semiotics: Studies on the emergence and evolution of human communication*. Amsterdam: John Benjamins.
Grandin, T., & Johnson, C. (2005). *Animals in translation: Using the mysteries of autism to decode animal behavior*. New York: Scribner Books.
Hoffmeyer, J. (2003). Semiotic aspects of biology: Biosemiotics. In R. Posner, K. Robering, & T. A. Sebeok (Eds.), *Semiotics: A handbook on the sign-theoretic foundations of nature and culture* (Vol. 3, pp. 2643–2666). Berlin: Walter de Gruyter.
Hoffmeyer, J. (2008). *Biosemiotics: An examination into the signs of life and the life of signs*. Scranton: Scranton University Press.
James, W. (1893). *The principles of psychology*. New York: H. Holt and Company.
Kashima, Y., & Haslam, N. (2007). Explanation and interpretation: An invitation to experimental semiotics. *Journal of Theoretical and Philosophical Psychology, 27/28*(2/1), 234–256.
Kull, K. (2010). Umwelt and modelling. In P. Cobley (Ed.), *The Routledge companion to semiotics* (pp. 43–56). London: Routledge.
Kull, K. (2011). Foundations for ecosemiotic deep ecology. In T. Peil (Ed.), *The space of culture – The place of nature in Estonia and beyond* (pp. 69–75). Tartu: Tartu University Press.
Kull, K. (2014). Zoosemiotics is the study of animal forms of knowing. *Semiotica, 198*, 47–60.
Kull, K. (2015). Semiosis stems from logical incompatibility in organic nature: Why biophysics does not see meaning, while biosemiotics does. *Progress in Biophysics and Molecular Biology, 119*(3), 616–621.
Kull, K. (2016). The biosemiotic concept of the species. *Biosemiotics, 9*(1), 61–71.
Kull, K., & Torop, P. (2011). Biotranslation: Translation between umwelten. In T. Maran, D. Martinelli, & A. Turovski (Eds.), *Readings in zoosemiotics* (pp. 411–425). Berlin: De Gruyter Mouton.
Magnani, L. (2015). Naturalizing logic: errors of reasoning vindicated: Logic reapproaches cognitive science. *Journal of Applied Logic, 13*(1), 13–36.
Magnus, R. (2011). Time-plans of the organisms: Jakob von Uexküll's explorations into the temporal constitution of living beings. *Sign Systems Studies, 39*(2/4), 37–57.
Maran, T., Martinelli, D., & Turovski, A. (Eds.). (2011). *Readings in zoosemiotics*. Berlin: De Gruyter Mouton.
Marler, P. (1978). Affective and symbolic meaning: Some zoosemiotic speculations. In T. A. Sebeok (Ed.), *Sight, sound, and sense* (pp. 113–123). Bloomington: Indiana University Press.
Martinelli, D. (2010). *A critical companion to zoosemiotics: People, paths, ideas*. Berlin: Springer.
Mendelson, T. C., Fitzpatrick, C. L., Hauber, M. E., Pence, C. H., Rodríguez, R. L., Safran, R. J., Stern, C. A., & Stevens, J. R. (2016). Cognitive phenotypes and the evolution of animal decisions. *Trends in Ecology & Evolution, 31*(11), 850–859.
Michon, J. A., Jackson, J. L., & Jorna, R. J. (2003). Semiotic aspects of psychology: Psychosemiotics. In R. Posner, K. Robering, & T. A. Sebeok (Eds.), *Semiotics: A handbook on the sign-theoretic foundations of nature and culture* (Vol. 3, pp. 2722–2758). Berlin: Walter de Gruyter.
Moore, B. R. (2004). The evolution of learning. *Biological Reviews of the Cambridge Philosophical Society, 79*(2), 301–335.
Müller, A., & Wolff, J. R. (2003). Semiotische Aspekte der Neurophysiologie: Neurosemiotik. In R. Posner, K. Robering, & T. A. Sebeok (Eds.), *Semiotics: A handbook on the sign-theoretic foundations of nature and culture* (Vol. 3, pp. 2667–2698). Berlin: Walter de Gruyter.
Nagel, T. (1974). What is it like to be a bat? *The Philosophical Review, 83*(4), 435–450.
Nemirovsky, P. (1999). *Aesthetic forms of expression as information delivery units*. MSc thesis, Massachusetts Institute of Technology
Nemirovsky, P., & Davenport, G. (2002). Aesthetic forms of expression as information delivery units. In P. Mc Kevitt, S. Ó Nualláin, & C. Mulvihill (Eds.), *Language, vision, and music:*

Selected papers from the 8th international workshop on the cognitive science of language processing, Galway, Ireland 1999 (pp. 255–270). Amsterdam: John Benjamins.

Pöppel, E. (1997). The brain's way to create "nowness". In H. Atmanspacher & E. Ruhnau (Eds.), *Time, temporality, now: Experiencing time and concepts of time in an interdisciplinary perspective* (pp. 107–120). Berlin: Springer-Verlag.

Pöppel, E., & Bao, Y. (2014). Temporal windows as a bridge from objective to subjective time. In V. Arstila & D. Lloyd (Eds.), *Subjective time: The philosophy, psychology, and neuroscience of temporality* (pp. 241–262). Cambridge: MIT Press.

Rattasepp, S., & Kull, K. (2016). The semiotic species: Deelying with animals in philosophy. *The American Journal of Semiotics, 32*(1/4), 35–48.

Rodríguez, H., Claudio, J., & Kull, K. (2017). The biosemiotic glossary project: The semiotic threshold. *Biosemiotics, 10*(1), 109–126.

Sebeok, T. A. (Ed.). (1968). *Animal communication: Techniques of study and results of research.* Bloomington: Indiana University Press.

Sebeok, T. A. (1972). *Perspectives in zoosemiotics.* The Hague: Mouton.

Sebeok, T. A. (Ed.). (1977). *How animals communicate.* Bloomington: Indiana University Press.

Sebeok, T. A. (1988). 'Animal' in biological and semiotic perspective. In T. Ingold (Ed.), *What is an animal?* (pp. 63–76). London: Unwin Hyman.

Sebeok, T. A. (1990). *Essays in zoosemiotics.* Toronto: Toronto Semiotic Circle.

Tulving, E. (2002). Chronesthesia: Conscious awareness of subjective time. In D. T. Stuss & R. T. Knight (Eds.), *Principles of frontal lobe functions* (pp. 311–325). New York: Oxford University Press.

Uexküll, J. von (1928). *Theoretische Biologie.* Berlin: Verlag von Julius Springer.

Uexküll, J. von (2010) [1934]. *A foray into the worlds of animals and humans, with a theory of meaning* (J. D. O'Neill, Trans.). Minneapolis: University of Minnesota Press.

Varela, F. J. (1999). The specious present: A neurophenomenology of time consciousness. In J. Petitot, F. J. Varela, B. Pachoud, & J.-M. Roy (Eds.), *Naturalizing phenomenology: Issues in contemporary phenomenology and cognitive science* (pp. 266–314). Stanford: Stanford University Press.

Vehkavaara, T. (1998). Extended concept of knowledge for evolutionary epistemology and for biosemiotics: Hierarchies of storage and subject of knowledge. In G. L. Farré & T. Oksala (Eds.), *Emergence, complexity, hierarchy* (pp. 207–216). Espoo: Organization. Finnish Academy of Technology.

Weber, A. (2016). *The biology of wonder: Aliveness, feeling and the metamorphosis of science.* Gabriola Island: New Society Publishers.

Of Men, Dogs and Bears: Communication in the Wilderness

Tarcisio Lancioni

Abstract This paper is about a story in which men and animals meet, in the deep of the wilderness, where nature is resisting culture. Although they, men and animals, do not share a language, or anything else that could be regarded as a system of conventional signs, nevertheless they communicate, revealing to us an idea of "community," and subsequently of "culture," that goes far beyond the common distinction between nature and culture.

It is of course fiction, but sometimes fiction, especially great fiction, represents a perfect laboratory which allows us to study not what or how things "are" but what they mean, to better understand the reason why things are what they are to us and how languages work to turn these things into what they are to us, whether we are men or animals.

1 Nature and Literature: A Few Words About the Universe

There is hardly a novel, short story, or poem that does not contain at least some image of nature, however diminished, denied, or distorted, presenting the reader with boxwood, ligustrum and acanthus, or even grassy ditches, tufts of reeds, and lemon trees[1]; lonely hillsides,[2] wuthering heights, wastelands, or impenetrable jun-

[1] Listen to me, the poets laureate
walk only among plants
with rare names: boxwood, ligustrum and acanthus.
But I like roads that lead to grassy
ditches where boys
scoop up a few starved
eels out of half-dry puddles:
paths that run along the banks,
come down among the tufted canes
and end in orchards, among the lemon trees.... (Montale 2012)

[2] Always to me beloved was this lonely hillside
And the hedgerow creeping over and always hiding... (Leopardi 1819)

T. Lancioni (✉)
University of Siena – Centro di Semiotica e teoria dell'immagine "Omar Calabrese", Siena, Italy
e-mail: tarcisio.lancioni@unisi.it

G. Marrone, D. Mangano (eds.), *Semiotics of Animals in Culture*, Biosemiotics 17, https://doi.org/10.1007/978-3-319-72992-3_11

gles; all these being simple backdrops for man's actions or objects of his productive and transformative work.

This pervasive presence of "nature" within literature could be attributed to man's need to represent himself in connection with his origins or his surrounding environment, which provides him with the necessary resources or inhibits his aspirations and undertakings, or even modulates and reflects his moods.

In this regard, I would like to present two different authors, who offer us their own interpretations and explanations of this almost obsessive presence of nature within our tales. The first is one of the greatest Italian poets, who is appearing for the defense:

> The entire world is made up of cells, which exists as different landscapes, each one animated by something that renders it unique with respect to any other landscape.
>
> There are therefore landscapes everywhere that deserve to be loved unconditionally. Because, in the end, the aim is also to faithfully represent the underlying reasons for our love for the Earth in its most original form. (*Andrea Zanzotto* (Interview with Marco Paolini). A film by Carlo Mazzacurati)

The second is a far more skeptical novelist, leading for the prosecution, who has perhaps grown a little bored, not to say nauseated, by the hyped-up "nature" presented by the romantics:

> We go to Nature for comfort in trouble, and sympathy in joy, only in books. Admiration of those beauties of the inanimate world, which modern poetry so largely and so eloquently describes, is not, even in the best of us, one of the original instincts of our nature. As children, we none of us possess it. No uninstructed man or woman possesses it. Those whose lives are most exclusively passed amid the ever-changing wonders of sea and land are also those who are most universally insensible to every aspect of Nature not directly associated with the human interest of their calling. Our capacity of appreciating the beauties of the earth we live on is, in truth, one of the civilized accomplishments, which we all learn as an Art […].
>
> What space do they ever occupy in the thousand little narratives of personal experience that pass every day by word of mouth from one of us to the other? [...] There is surely a reason for this want of inborn sympathy between the creature and the creation around it, a reason which may perhaps be found in the widely-differing destinies of man and his earthly sphere. (Collins 1865, p. 24)

Whether out of necessity or pure artifice, nature is omnipresent within literature, including the novels of Wilkie Collins, who dedicates long passages to describing wilderness or to pastoral imagery.

2 Forests

In order to somehow deal with this extensive pervasiveness and proliferation, rather than examining all the natural images in literature, I have selected one particular representation of which I have "fond literary memories," a broad selection of texts of various types and periods on the forest, the woods, or the wilderness. At which point I almost immediately realized that, rather than having circumscribed the object of my studies, I had merely switched from one order of infinity to another.

In fact, not only are novels and poems full of woods and forests, but so are figures of speech in literary texts, as these meta-literary figures are used to give form to the literature itself, if not to the thought itself.

In no particular order: Umberto Eco's walks in "Fictional Woods," Eliot's "Sacred Wood," and Baudelaire's (and Victor Turner's) "Forests of Symbols," not to mention the imaginative transformations, with literary repercussions, of *Holzwege* and *Lichtung*. Figures of speech that not only work well in narratives but also help us to understand literature and its implied imagery.

Furthermore, my purpose is not merely providing an account of the literary and meta-literary pervasiveness of forests, as a number of highly qualified studies in this respect already exist, viz., three Italian encyclopedia articles listed in the bibliography (Baffetti 2003; Ceserani et al. 2007). For certain historical aspects of the development and diffusion of the topic, on the other hand, reference is made to the "classic" works of Robert Curtius (1948) and Northrop Frye (1957).

The latter two both agree on the rarity of literary forests in antiquity (when they were, however, a much more prominent aspect of the landscape than in more recent times) and their subsequent proliferation from Medieval times to the age of Romanticism, pervading both folklore and popular literature, as confirmed by Vladimir Propp, who dedicated the entire first chapter of his *The Historical Roots of Fairy Tales* (1997) to the "forest."

All these forests are dark and intricate places (the daylight hardly penetrates any self-respecting forest, except for the occasional clearings) where outlaws and adventurers encounter marvels and monsters: this was the case with Chrétien de Troyes' Yvain and Ariosto's knights errant, who met with deception, glory, and madness in such places. And it was precisely during Medieval times and the Age of Chivalry that the fame of the great European forests was established, e.g., that of the Ardennes, of Broceliande, of Sherwood, and of the *Schwarzwald*, not to mention the heroes who populated these forests: Robin Hood, who appears across the entire range of popular and romantic English literature, from the Middle Ages until the time of Walter Scott, as well as the protagonists of numerous fairy tales, such as Thumbelina, Hansel and Gretel, and Snow White, furthermore the forerunners of the heroic and anti-heroic figures who penetrated the "Heart of Darkness" of the African forest (from William Burroughs to Amos Tutuola) and finally the characters of Joe Lansdale and Cormac McCarthy, who return to celebrate America's "lost forest."

I shall not dwell on the category of music, although forest-related topics emerge there as well, once again most frequently as "romantic" themes, from Von Weber and Wagner, to The Cure.

Neither is this the place to digress on the representations of forests in paintings, which are rarer still. There are of course painters who choose them as subjects in their works, but they are nevertheless few and far between: these include some seventeenth-century Dutch landscape painters, romantics such as Caspar David Friedrich, some French naturalists of the so-called Barbizon school, or a late impressionist like Cézanne, not to mention several American naturalist painters whose main interest was documenting the gradual conquest of the continent (see the Hudson river school), furthermore a number of informal artists like the Italian Morlotti, symbolists like Böcklin, Degouve de Nuncques, or Klimt ("Tannenwald"), but little else. Why this discrepancy? Why all this talk about it, and so few paintings? This is indeed strange, considering it is "an image."

Perhaps it is because forests are places of darkness, and it is difficult to bring out darkness in paintings, or perhaps, as Ortega Y Gasset (1914) sustains in the first of his *Meditations on Quixote*, it is because "the forest cannot be seen," thereby revealing the extraordinary truth behind the statement "You cannot see the forest for the trees." In the woods, what we see is trees, behind which we can only imagine there being more trees, or perhaps something else entirely, while being unable to grasp the depth of the mass of trees necessary to constitute "a forest." According to Ortega Y Gasset, the task of the manifest trees is to render the others absent, or merely hypothetical. Behind the trees we see we can imagine a forest, but we cannot see it. Thus there seems to be no difference between a boundless forest and a small patch of trees, at least visually.

It is only from the outside, and from a high-up vantage point (raised, and from a distance), that we can appreciate the expanse that transforms the trees into "a forest." But this would only be a look "from the outside," or more exactly a look that would reveal to us a compact mass, a "wall," like in certain works by Friedrich, and nothing of the "forest" itself, i.e., of what it might contain.

It is precisely this interplay between visibility and invisibility that renders a forest the ideal place of concealment; of the sudden appearance, or equally sudden disappearance, of fantastical figures, seductive and elusive nymphs, or ogres lying eternally in wait.

According to Jacques Le Goff, it is precisely this ability to conceal that turned the woods into the quintessential wilderness of medieval culture, as well as of the later romantic period. A place which is home to all those shunning civilization and culture, be they outlaws or hermits, and which hides all those things that can only exist far away from society, and under particular conditions, like apparitions and miracles.

In this sense, the forest is the perfect analogy to other spatial images, such as the sea with its islands and the desert with its oases, as noted by Le Goff himself:

> [In] the medieval West [...] the fundamental dualism between nature and culture presented itself predominantly as a contrast between what was built, cultivated, and inhabited (city, castle, and village) and what was essentially wild (the ocean, the forest: the western equivalents of the eastern desert). (Le Goff 1983, p. 44)

And moreover:

> Thus the forest was the equivalent of what the desert was in the East: a refuge, a hunting ground, a place of adventure, impenetrable to those who lived in cities and villages or worked the fields. (Id., p. 108)

The homology between these spaces, i.e., their interchangeability in certain narrative contexts, is made possible by the same figurative structure that articulates an opposition between a confined and definite space and an undifferentiated expanse, the latter characterized either by an excess of emptiness, or by a fullness which serves to isolate the confined space (the island, the oasis, the clearing, etc.), which is thus in turn rendered inaccessible or invisible on the horizontal plane, and therefore able to open up on the "vertical axis," upon which our relationship between the celestial world and the realm of chthonios is based. A necessary condition for the manifestation of the "supernatural."

All these types of places, which are structured around the same opposition between an undifferentiated expanse and a single incorporated and defined space, supply a "natural" place to "other" worlds, whose otherness cannot be expressed or imagined in other terms than as a reversal of our own world.

This oppositional structure is clearly expressed by Le Goff and Vidal-Naquet in an excellent structural analysis of *Yvain* by Chrétien de Troyes, dedicated to Lévi-Strauss, which shows how the insane Yvain's primitiveness is essentially a reversal of courtly life:

> This opposition manifests itself on two levels: between Yvain and the hermit, who is a "cultural" *enclave* within the "natural" world, and between Yvain and his ancient universe beyond the forest. Yvain chose the wilderness, namely the forest and its immediate elements: a system of clothing (torn clothes, final nudity), a nutritional code (processed and cooked food products replaced by raw foods), and a mental world (human memory is replaced by primary impetus and the repetitiveness of a wild existence); what Yvain has left behind is "culture", or rather an organized social system (agricultural production: cultivated fields, orchards protected by hedges symbolizing the boundaries of tilled lands); a system of habitation (tents, houses, pavilions) which is replaced by outdoor life, in the same manner that the hunting economy (with the help of a stolen bow) replaces the agricultural economy. (Le Goff and Vidal-Naquet 1973, p. 115)

As also noted by Le Goff and Vidal-Naquet, the lack of differentiation between these spaces, which can be summed up in their lack of culture, is not absolute but depends upon the structural system of relationships associated with the space itself. Thus the space might conceal particular methods of culturalization, which are simply different or even completely contrary to those of the world "outside" the forest. While these would still constitute images of a sort of "wilderness," at the same time, they represent the opportunity, the conditions of an "other" culture.

In this regard, quoting Marc Bloch, Le Goff and Vidal-Naquet first evoke the shady universe of the forest workers, which form an alternative "society" to that of the rural areas and small towns, followed by the "reverse" nature of the world that Yvain encounters in the heart of the forest. A "courtly" universe that nevertheless appears to be the opposite of that outside the forest and where the characteristics of the "normal" world beyond the forest are hyperbolically amplified.

The same structuring of the "forest space" is also presented in the cited Robert Pogue Harrison's book, namely, in the long chapter dedicated to the Italian philosopher Giambattista Vico. Contrary to what was later sustained by Curtius and Fry, Vico, in his *Scienza Nuova*, in 1744, argued that forests were more or less everywhere in the minds of the ancients, although often in a more figurative than direct form. For Vico, the forest is associated with "what comes before," thus positioning itself both against time and beyond the realm of time. It is thus identifiable with what has always resisted man and his efforts to transform the world.

According to Vico, after the flood these primeval forests began to be inhabited by the descendants of Noah, who in the meantime had lost their humanity. Solitary, wicked, and beastly creatures who turned into giants: "They were distrustful, brutal, restless, and incestuous, and did not have the slightest notion of a law which ranked above their instincts and passions" (Harrison 1993, p. 17), and in these conditions they roamed the dense forests seeing neither sky nor daylight, only thunder and lightning causing them to raise their eyes to the heavens. These wonders of light and sound were then attributed to an all-mighty creature, and for the first time an image formed in their vacant minds.

In order to worship that God, the giants began to create clearings from which they could contemplate the sky above them and within which they began to bury their dead, thereby creating a connection between the celestial world and the realm of chthonios, right there in the clearing itself, where the forest briefly opens up, instituting a "difference" that gives meaning to the senseless and infinite uniformity of the forest's expanse, as well as to life itself.

This initial contrast between the density of the dark and impenetrable forest and the bright clearing inserts a contrasting space within the forest itself, thereby playing with the polarity between the forest and that which lies outside, but on a different plane. And if what lies "outside" the forest is presented as an essentially human and civilized space, the space of the clearing, which also forms a contrast to the density of the forest, but from the inside, lends itself easily to inverse semantic operations, being a place of cultural "otherness" compared to the inhabited space outside the forest.

Furthermore, the exaltation of the vertical axis, resulting from the closure of the horizontal plane, renders the clearing a symbolically effective place for communicating with the celestial world as well as with the underworld.

Thus, the symbolic effectiveness of the forest does not seem to be an arbitrary choice but on the contrary appears to be motivated by its ability to pose itself as a perception-altering means and by its ability to express figurative semantic oppositions, such as "continuous" vs "discontinuous," "open" vs "closed," "full" vs "empty," and "visible" vs "invisible," which in turn are susceptible to further semantic impacts and to narrative expansions required to transform these same categories: i.e., those of appearance and disappearance, concealment and discovery, pursuit and flight, with all the tension implied by the constant conditions of limited visibility, and the uncertain knowledge that ensues (see also Lancioni 2009).

3 Nature and Culture

While I would like to illustrate how this "natural semiotic" device works within a story in which man and animal communicate, I should first like to reiterate that the forest had been selected as the exemplary image of "nature," namely, a wild form of nature, and thus as something entirely "other" compared to what is considered to be civilized or cultural.

As previously mentioned, this opposition between nature and culture, which seems so natural to us as to appear universal, has been questioned by several anthropologists, among whom Philippe Descola (2005), Eduardo Viveiros de Castro (2004; 2009), and Eduardo Kohn (2013), as well as by Bruno Latour (1999), a scholar of science with a particular interest in anthropological thinking, not to mention Gregory Bateson, who all reminds us that the issue is not merely a current one. I could also mention the work of Roy Wagner, whose *The Invention of Culture* (1975) already introduces many of the arguments that characterize Latour's thinking on this subject.

These anthropologists mainly insist upon the local and relative (entirely Western) nature of the contrast between nature and culture, and, in a manner that seems to me rather reductive, they tend to identify it with the contraposition human vs nonhuman (which also prevails in Latour's reflections), although it is evident that the various meanings of nature and culture cannot be generically reduced to this. For example, leaning on Hegel, Gadamer (1960) reminds us that in the eighteenth century the emerging concept of "culture" indicated the values of the community as opposed to the individual, and not as opposed to "nature."

Furthermore, although it is true that there is a marked difference between the Western conception of a man emerging from an animal substratum, maintaining this physical dimension while differentiating himself from it spiritually, and the Amazonian tribal conception of an animal emerging from a common human substratum while maintaining its spirituality and differentiating itself physically, this does not mean that any form of communicative continuity is impeded in the former.

Latour is certainly right when he argues that one must understand science in order to understand the West, but perhaps one risks falling into a symmetrical error if the "West" is equated with its laboratories.

The attribution of "subjectivity," especially if this subjectivity is understood (as it often explicitly is) in terms of the enunciational contrast between "person" and "nonperson," between "you and me" on the one hand and "he/she" on the other, does not coincide with that of "humanity," i.e., bearer of culture, as opposed to "nonhuman," i.e., relegated to nature, but the former attribution merely designates the relative positions of the actants within a communicative interaction.

I also get the impression that they tend to mix up two different categories of "nature.".[3] Resorting to a hjelmslevian image, which is more familiar to semioticians, we might say that we are dealing with a "nature" as *matter* and a "nature" as *substance*. In the first case, nature is viewed as the unique substratum in which various cultures can produce different images through a process of shaping.

And in the second case, we have a "nature" as *substance*, which does not consist of a single, uniform substratum but is rather the product of a particular cultural organization and valuation system. These are the broadly discussed naturalizations, the many "effects of nature" which now contribute to an apparently universal

[3] It is necessary, however, to acknowledge Descola's awareness of this fact when he cites Spinoza's distinction between *natura naturata* and *natura naturans* (Descola 2011).

consensus (Marrone 2011) and which appear as different images of this entirely cultural "nature-building" process.

Finally, I wonder to what extent the notion conceived by Lévi-Strauss, who was mainly responsible for rendering the contrast between nature and culture absolute, at least in De Castro's view, differed from this multifaceted view of "nature."[4] Although interested in classification systems, Lévi-Strauss was always highly involved in the reconstruction of local encyclopedias (*emic*), not in order to compare them with our knowledge of botany, astronomy, or zoology but on the contrary to use them as a reference framework for interpretive work. Furthermore, in my view it was Lévi-Strauss who taught us to see structures within myths and rituals, through which we are able to contemplate and resolve the complex relationships generated by the excessive proximity or distance between the various figures that populate a certain world.[5]

His writings do not seem to contain a strong defense of a single "nature," apart from that of the psychical work of man, which he viewed as undoubtedly universal, "natural," and responsible for a universal form of contrastive organization of knowledge (a well-known idea derived from structural linguistics). Otherwise, far from being naïve as some sustain, his writings do not convey a view of "nature" as the unchanging and identical substratum to every culture (the "nature" as *matter*), but "nature" is defined as that which each individual culture, each in its own way, pushes away from itself, without any ontological meaning. This meaning, which was later taken up by Greimas (1970), is based on the contrast appropriation/expulsion, which appears much more general than the psychological contrast between interiority and exteriority, as proposed by Philippe Descola.

The work of Eduardo Kohn is of particular interest to our discussion, starting with the title of his monograph on the Runa people: *How Forests Think* (Kohn 2013).

Kohn's work is explicitly linked to that of Descola and De Castro, whom he wants to take even further down the road toward an ethnology that goes beyond man, in the belief that it is essential to expose the ecological network to which man himself belongs in order to understand him. And he argues that semiotics is the only way to do so.

In short, Kohn believes that anthropologists have been incorrect in identifying "cultural" and "symbolic" (as defined by Peirce). In fact, "symbolic" does not exhaust the field of semiotics, which also includes "iconic" and "indexical," something which anthropologists have never bothered with. The latter two do as a matter of fact characterize the primary forms through which every living being

---

[4] For the sake of clarity, I would like to make reference the two texts of the so-called *clam* controversy with Marvin Harris, "Structuralism and ecology" and "Structuralism and empiricism", both of which are now contained in Lévi-Strauss 1983. Texts also annotated by Descola 2011.

[5] I will limit myself to noting "The story of Asdiwal", and how the rituals associated with seafood processing were seen as a means for redefining the distances between the human and salmon "brothers", without which every salmon meal would have been viewed as an act of cannibalism. (Lévi-Strauss 1973).

creates representations of its own environment, which thus becomes an immense network of semiotic interactions, in which each "subject" creates representations of its own environment, and of himself inside this environment, using their own semiotic means.

In the forest, socialization does not concern only human bonds, thus placing humans in *complete* contrast to nonhumans, but unites all beings who present themselves as bearers of "subjectivity," capable of presenting themselves to others as an "I" or a "you," and creating representations of the "other" and the environment.

I have limited myself to discussing these few aspects of Kohn's work in order to emphasize:

(a) How once again the elimination of a hypothetical "universal" (nature/culture) results in the assumption of other universals, which in this case are specifically semiotic: A sign-based typology, such as Peirce's, and an enunciational model (I/You vs He/It).
(b) The significance attributed to the forest environment, since it is only within this ecologically complex context that forms of widespread semiotic socialization would emerge, thus giving rise to a complex ecological (and not only human) semiotic system, supposedly completely "other" compared to our civilized world, and not in any way connected with this civilized view which apparently ontologically opposes humans and nonhumans, rendering them objects of separate knowledge.

4 The Bear

Does Kohn's argument really only apply to the world of the forest and only to "non-Western" cultures? We will see that the attribution of subjectivity to "nonhumans," and sometimes, more dramatically, the denial of subjectivity to humans themselves, is a common and widespread experience that does not necessarily undermine the nature vs culture opposition; however, it continuously redefines the extent and the scope of what we refer to as "culture," that is, what a society incorporates in its system of relations, albeit in varying degrees, and the tools employed to ensure this incorporation.[6]

Even in "our" world, the contrast between nature and culture is not without its nuances and does not necessarily separate humans from nonhumans.

This is the specific aspect that I would ultimately like to examine in greater detail, by looking more closely at one of our many literary forests: the *Big Woods* by William Faulkner and the American *Wilderness* as celebrated in one of his most famous short stories—*The Bear*.

Rather than a complete analysis of the short story, I would like to present a few notes pertaining to the subject matter in question.

---

[6] Cfr. Lancioni 2015.

The Bear is the story of young Ike McCaslin's initiation rites into hunting in the forest. From the very beginning, the story introduces as a motif the gradual disappearance of the forest, and with it the old ways of "being a man."

Thus a contrast is created between two worlds: "the city" on the one hand, characterized by the increasingly accelerating pace of progress, reaching out via a rail network that is entangling the entire subcontinent of the Mississippi like a cobweb, making its "golden voice" of money and power heard; and on the other hand *The Wilderness*, exemplified in the shape of the big forest, which has been there forever, even before it was inhabited by the first Indians, and before the continuous waves of white people attempted to take possession of it. A world that resists the sterile activity of "countless men, unknown even to each other," attempting to chip away at its edges:

> The same solitude, the same loneliness through which frail and timorous man had merely passed without altering it, leaving no mark nor scar, which looked exactly as it must have looked when the first ancestor of Sam Father's Chickasaw predecessors crept into it and looked about him, club or stone axe or bone arrow drawn and ready. (p. 30).[7]

A place whose history is only marked by the memory of the few who were able to leave a lasting impression through their endeavors, whether wild beasts or bandits, whose echoes continue to haunt the forest in the form of voices recounting their tales.

The pace of the forest is therefore not one of progress, oriented toward the future, but is rather one of "legendary" permanency, annihilating the flow of time by lining up these timeless notorious figures next to one another, without a before or after.

Although the *Big woods* where the events of the story take place have since become a "reserve," property of Major De Spain, and therefore already "entangled" in the cultural dimension of "law and order" as well as in private property, they still seem capable of instilling a sort of "suspension" of this reality, where "other laws" that belong to a different "world order" continue to apply.

The polarity between the two worlds of the *city* and the *wilderness* is not that of "nature" vs "culture," but rather one between two distinct "natures and cultures": on the one side, the culture of the city, made up of institutions and money, and whose "nature" is an economic resource to be looted, risking destruction in the absence of any respect whatsoever for the forest, its trees, and its animals. This vision triumphs at the end of the story, when the large sawmills linked up by an increasingly dense network of railways replace the forest landscape.

On the other side, the culture of the wilderness, made up of rules of honor that everyone must respect in order "to be somebody," "to be a man." This goes for the men themselves, who are called upon to be such regardless of race or social status (distinctions that are essential elsewhere, above all in Faulkner's universe), as well as for the animals, be they dogs (which are not merely helpers "programmed" to flush out their prey, but called upon to distinguish themselves as either clever or stupid, brave or cowardly), or "noble" beasts, big game like bear or deer, which are

[7] All quotations from Faulkner 1955.

expected to "obey the rules": all are subject to a system of interaction, unchanged by time, in which they all can acquire their own subjectivity, their own individuality, an "I," only if they show humility and tenacity. A world fueled and kept alive by stories, recounted around an open fire, with a glass of whiskey:

> It was of the men, not white nor black nor red but men, hunters, with the will and hardihood to endure and the humility and skill to survive, and the dogs and the bear and deer juxtaposed and reliefed against it, ordered and compelled by and within the wilderness in the ancient and unremitting contest according to the ancient and immitigable rules (pp. 14–15)

A world, that of the forest, only populated by men and animals, without women, which can only be accessed by ritually shedding the "civilization" of the other world. This separation is perceivable from the moment of access, which marks the transfer to a place where it is difficult to find one's bearings, where a different perception is imposed, determined by conditions which only allow for a short range of vision, in faint light, and where vision is replaced by intra-vision:

> The wilderness closed behind his entrance as it had opened momentarily to accept him, opening before his advancement as it closed behind his progress, no fixed path the wagon followed but a channel nonexistent ten yards ahead of it and ceasing to exist ten yards after it had passed, the wagon progressing not by its own volition but by attrition of their intact yet fluid circumambience, drowsing, earless, almost lightless. (p. 20)

A separation that is also marked by a series of transformations not unlike those Le Goff finds in the *Yvain*: nutritional, as sour bread and venison are the only foods allowed; with regard to roles, as only men ritually cook the food; and with regard to "comfort," as the soft sheets of the city are replaced by rough blankets. It is not so much an "abandonment" of civilization as it is a "reduction" of it, man approaching the other inhabitants of the forest. This approach threatens to undermine the usual classification system, and not only in terms of the strict categories that distinguish and separate men from each other, as we see in the opening lines of the story, which actually begins with a classification exercise:

> There was a man and a dog too this time. Two beasts, counting Old Ben, the bear, and two men, counting Boon Hogganbeck, in whom some of the same blood ran which ran in Sam Fathers, even though Boon's was a plebeian strain of it and only Sam and Old Ben and the mongrel Lion were taintless and incorruptible. (p. 14)

No matter how we twist and turn it, the classification does not hold: no hypothesis seems capable of reconciling the world presented here with the taxonomic organization of "our world," the world of the city, regardless of how our city may differ from the one Faulkner presents. Is the bear Ben not more of a beast? And is Boon not more of a man? And who is the "other" man? There could be several in the story. The boy? Whose initiation and whose encounter with the bear constitute the main plot? But he is just a 10 year-old kid when the story begins, and 16 when it ends. Or what about Sam Fathers, named to indicate class and race relationships in relation to Boon? Class and race which, as previously mentioned, lose their meaning in the forest, to the point that, as we are told, only two half-breeds, a man and a dog (Lion), along with Ben, can be considered "pure." And if purity constitutes full membership of a class, as Mary Douglas suggests, what category can then only include Ben,

Lion, and Sam? Perhaps it is their ability to resist change, and their refusal to "connect" to the civilization, which is about to engulf the world of the forest, after which they will perish too. A resistance that renders them isolated and lonely figures: Sam lives isolated from other men, and more than anyone else, he belongs to an "ancient" world, because he is the descendant of an Indian Chief ("the savage who had come out of the forest less than a generation ago, with no children, no family and no people," p. 71). Lion too, the indomitable and untamable dog captured by Sam and adopted by Boon, is indifferent to everything and has no bonds. And finally there is the bear, the solitary and legendary monster. A true Moby Dick of the forest. In Faulkner's words, the epitome and apotheosis of an ancient world doomed to extinction.

Old Ben is one of those figures who are able to leave their mark on the life of the forest. A timeless and lawless creature, it too entirely alone:

> Not even a mortal beast but an anachronism indomitable and invincible out of an old dead time, a phantom, epitome and apotheosis of the old wild life which the little puny humans swarmed and hacked at in a fury of abhorrence and fear like pygmies about the ankles of a drowsing elephant;—the older bear, solitary, indomitable, and alone; widowered childless and absolved of mortality—old Priam reft of his old wife and outlived all his sons. (p. 17)

A figure that both metaphorically and metonymically belongs to the forest itself, against whom tiny anonymous men pit themselves to no avail,[8] leaving only scars and disfigurements, like the crippled foot that leaves its unmistakable tracks:

So legendary that "he" has earned his own name, like a man: Old Ben.

> [...] he would hear them still talking, about the shoats and calves Old Ben had killed and the cribs he had rifled and the traps and deadfalls he had wrecked and the lead he probably carried under his hide—Old Ben, the two-toed bear in a land where bears with trap-ruined feet had been called Two-Toe or Three-Toe or Cripple-Foot for fifty years, only Old Ben was an extra bear (the head bear, General Compson called him) and so had earned a name such as a human man could have worn and not been sorry. (pp. 68–69)

> [He] had earned for himself a name, a definite designation like a living man. (p. 16)

Furthermore, Old Ben is presented not only as an "he" instead of an "it" but also as an "I," as Kohn would have said, in every respect: the bearer of a subjectivity that not only translates into the ability to hunt and escape but also into the ability to organize the world through "representations" on the basis of which specific agentive programs (narrative programs in Greimasian terminology) can be determined (Greimas 1974). I shall cite a few examples, which I think are particularly relevant.

During their first stay at the hunting camp, the bear's tracks are found near the encampment, and Sam explains to Ike that the bear visits the camp each year, and people say that it does so to warn the young bears of the danger posed by the hunters, but this is not the real reason, Sam says:

---

[8] Who in turn are reinforced by the packs of hunting dogs, which seek in vain to show their courage confronting the bear.

> He come to see who's here, who's new in camp this year, whether he can shoot or not, can stay or not. Whether we got the dog yet that can bay and hold him until a man gets there with a gun. Because he's the head bear. He's the man. (p. 24)

Therefore, not only does it have a name like a man, and acts like a man would, but "he," literally, "is the man."

For years, Ike only senses the bear's presence around him indirectly, like when, during the sudden silence when lying in wait, he smells the "odor" and knows that Ben is there watching him, without ever allowing himself to be seen. One summer, when a bear hunt was not on the agenda, Ike decides to meet Ben and prepares a ritual for approaching him. He starts by going out, from dawn to dusk, alone, without a rifle, taking only a stick, a compass, and a watch to find his way in space and time. He finds the bear's tracks, but never encounters it; until one day he realizes that leaving the rifle behind is not enough. He must go even further, become even less "civilized," and more of a forest creature, if he wants the bear to accept him. He has to "purify himself."

> He had already relinquished, of his will, because of his need, in humility and peace and without regret, yet apparently that had not been enough, the leaving of the gun was not enough. He stood for a moment—a child, alien and lost in the green and soaring gloom of the markless wilderness. Then he relinquished completely to it. It was the watch and the compass. He was still tainted. He removed the linked chain of the one and the looped thong of the other from his overalls and hung them on a bush and leaned the stick beside them and entered it. (p. 35)

Ike gets completely lost, as if the forest was not really a forest until his disorientation was complete. Only then could the prodigious encounter take place: Ike recognizes a fresh pair of bear tracks, then another, and another, seeing them one at a time just as they are about to vanish, from an extremely close and "short-sighted" perspective in which there is no room for an orientation plan. Then, as he enters a clearing, he first notices the objects that he had abandoned, after which he sees the bear:

> […] emerging suddenly into a little glade and the wilderness coalesced. It rushed, soundless, and solidified—the tree, the bush, the compass and the watch glinting where a ray of sunlight touched them. Then he saw the bear. It did not emerge, appear: it was just there, immobile, fixed in the green and windless noon's hot dappling, not as big as he had dreamed it but as big as he had expected, bigger, dimensionless against the dappled obscurity, looking at him. Then it moved. It crossed the glade without haste, walking for an instant into the sun's full glare and out of it, and stopped again and looked back at him across one shoulder. Then it was gone. It didn't walk into the woods. It faded, sank back into the wilderness without motion as he had watched a fish. (p. 39)

In order to be effective, the "shedding" that allows the boy to be "accepted" by the bear requires a shared "code": not only the boy himself but also the bear must understand the meaning of the gesture, and both need to interpret it in the same manner. They must have the same ability to interpret this symbolic behavior. The bear must envisage an accurate representation of the boy's gesture and intentions to itself, and the boy must do the same in relation to the bear. To both, this "game" of interaction is not one of hunting, predation, or sacrifice, but it is rather a "presentation," a purely communicative exchange with no other practical purpose; the search for "affinity" or, as Landowski (2005) would say, for adjustment. The two subjectivities that constitute and recognize themselves in this way are not metaphorical subjectivities but are rather, as Goodman would say, literal ones. We are dealing with the creation of an actual space of real interaction, whose existence implies, and denotes, that complete subjectivity which the interlocutors attribute each other.

The forest is the bear's territory, and the boy, being an outsider, is the intruder and must learn the rules in order to be recognized as an "acceptable" fellow creature in this world. Following which he will be permitted to interact with the other inhabitants of this world, who, like Old Ben, possess subjectivity, whereas mediation and interaction is impossible with creatures deprived of subjectivity, such as snakes that infest the undergrowth.

Thus, while on the one hand the forest constitutes a place of complete otherness with respect to the city, the traces of which one must eliminate in order to be accepted, on the other hand, it does not in any way present itself as a shapeless *hyle*, as "nature" (matter) in opposition to "culture," but on the contrary as the place of another culture. According to Lotman (Lotman and Uspenskij 1973), we might say that it is the space of an anti-culture.

In this regard, it should be noted that the boy's initiation does not intend to integrate him into the "civil" society of the city, which would establish the forest as the heterotopic space for the acquisition of competence, but instead its purpose is to acquire a dimension which is "other" compared to that of civilization, and where "men" are characterized by humility and tenacity, regardless of race, class, or species. A humanity that likens Ike to Sam, but also to Old Ben, and which poses itself as an alternative to that of the "city."

The "ritual" hunt constitutes an alternative school to that of the city, which, with its stale ways, only serves to create anonymous social roles in the name of progress.

The hunt, the forest, and the bear itself are all part of a school with its own separate knowledge, places, and teachers, and with its own educational pathway:

> If Sam Fathers had been his mentor and the backyard rabbits and squirrels his kindergarten, then the wilderness the old bear ran was his college and the old male bear itself, so long unwifed and childless as to have become its own ungendered progenitor, was his alma mater. (p. 41)

Old Ben, then, is not just *like* a man, but he can even be the kid's alma mater, teaching him how to become "a real man."

We are inside a tale, of course, and in tales, as in fables and in cartoons, we know, it is not so strange to meet animals that act, think, and talk just "like men." But we easily feel that in The Bear, we are not in this kind of fictional anthropomorphism. Rather like in Kohn's Amazonian forest, we perceive that in Faulkner's wilderness men and animals, or more to the point some men and some animals, really communicate, and this happens because they interpret each other in a shared semiotic environment. A semiotic environment made up of ritualized gestures, spatial codes, body language, and so on that define "a culture," in which some men and some animals are included, and they are included only because they share this semiotic environment, only because they recognize themselves as "I" and "you," because they communicate, while other men and other animals are excluded, becoming "nature." Thus, the contrast is not between culture and nature, but rather between cultures and natures, their distinction depending on the constitution of semiotic environments.

But could this happen only in the "wilderness," far away from our "cities," as both Faulkner and Kohn seem to suggest? I think that humans and nonhumans do not engage in communicative interactions only in remote wilderness, or only in fables and cartoons, but also in ordinary everyday situations, as is well known to all those who love spending time in prayer, and to those who interact with their pets on a daily basis. I do not refer to when we "pretend" to talk with them, but rather to when we try to actually interact with them, starting with eye contact. This is no metaphorical form of communication, since an actual space of interaction is created, in which we make room for the other party so that they can enter into contact with us, becoming a "you," and we expect that the other party will make room for us as well, turning us into an opposite "you".

One more brief note to finish this analytical sketch, which I would have liked to develop further. In the version contained in *The Big Woods*, as well as in the one contained in *Go Down Moses* (Faulkner 1940) (although here the dominant narrative concerns the complex racial plots and the ensuing dramas), *The Bear* presents us with a forest to be taken as an exemplary image of the *Wilderness*, which stands in contrast to the world of the city.

In *Collected Stories*, there is a section with the title *The Wilderness*. However, it does not include forest at all; on the contrary it only contains tales dealing with legal cases and courtrooms, which are consequently seen as images of a "wild" world.

Once again, just like at the beginning of "our" story, the classification game makes fun of our taxonomies and expectations.

But perhaps that is precisely why the story is worth telling. Maybe that is what novels and stories are for, to make us reconsider the "shape" of the world and its manifestations and to tell us that nature and culture, just as life and death, are in constant motion and that they are expressions of numerous and shifting ways of conceiving the world around us. And if rethinking the world order is a good reason to tell a story, then Faulkner is certainly a great storyteller. But this, of course, we knew already.

References

Baffetti, G. (2003). Foresta. In Anselmi, G.M. and Ruozzi, G. (eds). *I luoghi della letteratura italiana*. Milano: Bruno Mondadori 2003.

Ceserani, R., Domenichelli, M., & Fasano, P. (Eds.). (2007). *Dizionario UTET dei temi letterari*. Torino: Utet.

Collins, W. (1865). *The woman in white*. New York: Harper & Brothers.

Curtius, E. R. (1948). *Europäische Literatur und lateinisches Mittelalter*. Bern: Francke Verlag.

De Castro, V. (2004). *Exchanging perspectives. The transformation of objects into subjects in Amerindian ontologies*. Durham: Duke University Press.

De Castro, V. (2009). *Métaphysiques cannibales. Lignes d'anthropologie post-structurale*. Paris: PUF.

Descola, P. (2005). *Par-delà nature et culture*. Paris: Gallimard.

Descola, P. (2011). *L'écologie des autres. L'anthropologie et la question de la nature*. Paris: Quae.

Faulkner, W. (1940). *Go down Moses*. New York: Random House.

Faulkner, W. (1955). *Big woods*. New York: Random House.

Frye, N. (1957). *Anatomy of criticism. Four essays*. Princeton: Princeton University Press.

Gadamer, H. G. (1960). *Wahrheit und Methode*. Tübingen: J.C.B. Mohr.

Greimas, A. J. (1970). *Du sens*. Paris: Seuil.

Greimas, A. J. (1974). *Du sens II*. Paris: Seuil.

Harrison, R. P. (1993). *Forests. The shadow of civilization*. Chicago: University Of Chicago Press.

Kohn, E. (2013). *How forests think. Toward an anthropology beyond the human*. Oakland: University of California Press.

Lancioni, T. (2009). *Immagini narrate. Semiotica figurativa e testo letterario*. Milano: Mondadori Università.

Lancioni, T. (2015). Appareils de capture. Pour une sémiotique de la culture. *Actes Sémiotiques*, 118.

Landowski, E. (2005). *Les interactions risquées*. Limoges: Pulim.

Latour, B. (1999). *Politiques de la nature. Comment faire entrer les sciences en démocratie*. Paris: La Découverte.

Le Goff, J. (1983). *Il meraviglioso e il quotidiano nell'Occidente medievale*. Bari: Laterza.

Le Goff, J., & Vidal-Naquet, P.(1973). Lévi-Strauss en Brocéliande. Esquisse pour une analyse d'ur roman courtois. In AAVV, *Lévi-Strauss*. Paris: Gallimard.

Leopardi, G. (1819). The infinite. *Listener*, 43, no. 1113.

Lévi-Strauss, C. (1973). *Anthropologie structurale deux*. Paris: Plon.

Lévi-Strauss, C. (1983). *Le regard éloigne*. Paris: Plon.

Lotman, J., & Uspenskij, B. (1973). *Tipologia della cultura*. Milano: Bompiani.

Marrone, G. (2011). *Addio alla natura*. Torino: Einaudi.

Montale, E. (2012). The lemon trees. In: *id. Collected poems, 1920–1954*. New York: Farrar Straus & Giroux.

Ortega y Gasset, J. (1914). *Meditaciones del Quijote*. Madrid: Residencia de Estudiantes.

Propp, V. (1997). *The historical roots of fairy Tales*. Washington: Western Washington University Press.

Wagner, R. (1975). *The invention of culture*. Chicago: University of Chicago.

The Mixed Category Human-Animal in New Anthropology and in the Arts

Tiziana Migliore

Abstract Considering animality in terms of interdependency between humans and animals may help us understand how different species have evolved and continue to do so through time. The stakes are high. Darwinian evolutionary theories, while based on the continuity of the species and while suggesting the idea of mutual derivation, e.g. of humans from animals, are mostly concerned with the origins; they are not concerned with whether and how humans and animals have evolved together socially and culturally. Even a superficial look at contemporary society, with its habits, tastes and recurring trends in fashion, design and the arts, shows that the border between the two categories is nothing but a threshold: the *Umwelt* of the humans has evolved by borrowing from the *Umwelt* of the animals, creating hybrids and interspecies (Ramos, Animals. MIT Press, Cambridge, MA, 2016).

The article presents a few illustrations of the forms and force of this evolution, and highlights the obstacles that may prevent various disciplines from seeing it as it is, i.e. as a participatory process.

1 Mimicking the Animal

Semiotic research on mimicry and camouflage (Migliore 2008; Fabbri 2008, 2011; Casarin and Fornari 2010) tends to overlook a crucial fact: the chameleon's strategy shows not only that subjectivity is the result of a complex system of presentation and representation simulacra (of the self and of the other) (Fabbri 2008, p. 89) but also how a species may abandon its own physiological identity to adopt the physical and outer identity of another realm. The animal mimics the *facies* and the actorial, spatial and temporal characteristics of the environment in which it plans to hide; the soldier mimics the *facies* and the actorial, spatial and temporal characteristics – i.e. features, movements and behaviours – of animals that are good at hiding. The animal selects and adopts, that is, *embodies*, the most efficient defensive patterns of its own habitat; human beings select and *embody* the most efficient defensive and

T. Migliore (✉)
University Ca' Foscari, Venezia, Italy
e-mail: tiziana.migliore@unive.it

G. Marrone, D. Mangano (eds.), *Semiotics of Animals in Culture*,
Biosemiotics 17, https://doi.org/10.1007/978-3-319-72992-3_12

aggressive patterns of the animal. The oppositions human/environment and human/animal are not binary but *participatory*.

But do humans simulate these features in the sense that they are lying? Not at all. They shed their skin and adapt so well to their environment that some of the characters, initially learned out of contingency, eventually become stable, "natural" and necessary. The semiosphere remains the same. The evolution of the species, which unfortunately does not coincide with their progress, is determined by the dissemination and consolidation of these mimicry strategies. Historically, studies on animal mimicry, such as that of Henry Walter Bates, have had a crucial impact on the hypotheses of Wallace and Darwin.[1] René Thom provides extensive explanations of how such morphological mutations unfolded in time, going so far as to suggest that the shape of the predator's claw or beak may have adapted to the gradual increase or decrease in the size of the prey. Skin and shell are tactical interruptions of the prey's organic extension and counter the obstructive action of the predator's beak, fangs or claws (Thom 1988, p. 139). The highly sophisticated survival strategies of animals (Caillois 1960; Belpoliti 2015), often overlooked, could help us understand several human behavioural patterns, whether at war, during espionage operations or, more covertly, in everyday life.

Shall we then say, by extension, that *homo homini lupus*? If the claim may sound too extreme, it is a fact that the survivors are those who know how to run like leopards or slither like snakes, not those who move backwards like shrimps – "*a passo di gambero*", as suggested in the title of Umberto Eco's eponymous book (Eco 2006). The model of the human body, both at war and in everyday life, is based on the corporeality of animals.

These facts find no correspondence in the reflections on corporeality developed by speculative disciplines, raising the urgent issue of bridging this gap between experience and knowledge.

2 The "Corpus Archetypus"

At least since Polykleitos' *Canon*, Western culture has been dominated by a univocal notion of the body – anthropomorphic, self-standing and serving as the "*Nullpunkt of all* the dimensions of the world" (Merleau-Ponty 1964, p. 249). This point has been demonstrated for more than half a century by art historians focussing on perspective with a central vanishing point. In philosophy, this particular way of thinking about the body underlies the principle of subjectivity. Admittedly, twentieth-century thinkers did deconstruct the notion of the body as an enclosed and impervious referentiality, prior to sense, and insisted on aesthesia, including its non-codified and non-codifiable expressions, as the medium whereby the flesh and skin of corporeality, viewing and visible, partake in the world (Fabbri 1991; Fontanille

[1] See H. W. Bates, Letter to C. R. Darwin, 2 May 1863, in Darwin Correspondence Project, "Letter no. 4138", accessed on 24 March 2017, http://www.darwinproject.ac.uk/DCP-LETT-4138

2004; Marrone 2005). In phenomenology, however, the scenario remains that of a subject with a purpose and grasp on objects, with protension and retention – the subject of Husserl's intentional act.

Thus, the theory of embodiment or of the "embodied mind", which locates cognition in the body and its sensorimotor processes (Varela et al. 1991; Lakoff and Johnson 1999), and its derived theory of the "extended mind", which suggests that the environment and environmental interaction play a constitutive role in the formation of mental processes (Clark and Chalmers 1998; Clark 2008; Wilson 2004), seriously challenged what Isabelle Stengers calls the greatest "partition" in Western thought and the greatest obstacle to the progress of knowledge: the Cartesian dualism between mind and body (Stengers 1997). Both "embodiment" and "extended mind" theories criticise the ideal of an archetypal intellect that produces everything and in which everything is produced, and argue that cognitive and mental processes are expressed in the body (embodiment) and in the environment (extended mind). The problem, however, is that these new theories have changed our notions of both mind and the environment without rethinking our notions of the body, as if the latter were a universal principle, something given a priori with a hypostatised form. The "body", which is at the core of the model of the embodied mind, seems in no need of any definition.

As a consequence, the main theorists of the "embodied mind" struggle to explain how the mind is actually embodied and why neurons are not sufficient:

> Every living being categorizes. Even the amoeba categorizes the things it encounters into food or nonfood, what it moves toward or moves away from. The amoeba cannot choose whether to categorize; it just does. The same is true at every level of the animal world. Animals categorize food, predators, possible mates, members of their own species, and so on. How animals categorize depends upon their sensing apparatus and their ability to move themselves and to manipulate objects. Categorization is therefore a consequence of *how we are embodied*. We have evolved to categorize; if we hadn't, we would not have survived. Categorization is, for the most part, not a product of conscious reasoning. We categorize *as we do* because *we have the brains and bodies we have* and because we interact in the world *the way we do*. (Lakoff and Johnson 1999, pp. 17–18; italics mine)

In the next passage, Lakoff and Johnson return to the role of neurons and synaptic connections, since, in their view, "A small percentage of our categories have been formed by conscious acts of categorization, but most are formed automatically and unconsciously as a result of functioning in the world", so that "we do not, and cannot, have full conscious control over how we categorize" (ibidem, p. 18). And below:

> Think of the properties of the human body that contribute to the peculiarities of our conceptual system. We have eyes and ears, arms and legs that work *in certain very definite ways* and not in others. We have a visual system, with topographic maps and orientation-sensitive cells that provides structure for our ability to conceptualize spatial relations. Our abilities to move *in the ways we do* and to track the motion of other things give motion a major role in our conceptual system. The fact that we have muscles and use them to apply force *in certain ways* leads to the structure of our system of causal concepts. What is important is not just that we have bodies and that thought is *somehow* embodied. What is important is that the *peculiar nature* of our bodies shapes our very possibilities for conceptualization and categorization. (ibidem, p. 18–19, italics mine)

But how do we do this? Lakoff and Johnson's passage is relevant insofar as the capacity to categorise is attributed to both humans and animals and considered as an effect of their own embodiment. The category of the body, however, is constantly and openly treated as a given. According to the two authors, the "embodied mind" has no repercussion on the conception of the "body", which remains a priori: embodiment theories notwithstanding, "we have the brains and bodies we have" (ibidem, p. 18). Unlike evolutionary scientists, Lakoff and Johnson do consider how social and cultural coevolution may have unfolded since the origins. However, their hypotheses about categorisations do not even consider the possibility that the "body" may evolve in parallel.

Thus, we find that we have successfully overcome the myth of *intellectus archetypus*, but not the premise of the *corpus archetypus*, although the latter corresponds to the body on which the former is based. An approach so detached from the description of processes and practices inevitably overlooks the dynamic relationship between *sema* and *soma*, as well as the actualisation of the meanings through usage. Semiotics, the science that studies the life of signs inside social life (Saussure), must act as an intermediary between philosophical thought and concrete experiences. The semiologist must know how to become a "serial describer" (Paolo Fabbri).

3 Interdependent Physicalities

If we want to demonstrate how humans, through time, have observed and assimilated a number of behaviours borrowed from animal "societies", and hence the participatory nature of evolution, we need to debunk the myth of the single-body model. We should account, on the one hand, for the animal physical entity, and on the other, for the possibility of human-animal physical entities that are not separate but discontinuous – a realm in which evolution does not work by filiation but through the "cross-breeding" of heterogeneous terms. To begin with, one should note that humans tend to judge and ratify themselves on the basis of animal features. The examples are countless: fables with animal protagonists, expressions such as the above-mentioned "run like a leopard" or "slither like a snake", proverbs (*homo homini lupus*, "curiosity killed the cat", "pigs get fat, hogs get slaughtered"), the bestiaries that are at the origins of fabulous geographies and symbolic and hieroglyphical conceptions of the world and, finally, the genre of the caricature, which draws on the "dogma of physiognomy" that is on the "pseudo-Aristotelian idea according to which the human character can be determined from the similarity of the human countenance to that of certain animals" (Gombrich and Kris 1938, p. 319). This is quite the opposite of saying that humans project their own feelings and desires on animals. To the contrary, it is the animal world that offers organisational hints to humans. Thus, in caricature, "the unlike parts are united to give a new meaning. They form an animal in which the portrait of man lives on" (ibid.), so that "*nel tutto appariscono essere essi [le persone ritratte] e nelle parti sieno variati*" ("the portrayed subjects appear the same in the whole, but different in the single parts" Baldinucci 1681, entry "*caricare*", p. 111).

3.1 *The Human/Animal Participatory Category: Detour and Access*

What is the best approach to define the interdependency between humans and animals? Deleuze and Guattari's notion of "becoming-animal" (Deleuze and Guattari 1980) remains anchored in a human perspective, despite its pivotal contribution to twentieth-century Western philosophy and its challenge to the assumptions of unchanging humanity and unbridgeable gap between *Homo sapiens* and "world-poor" animals (Heidegger 1929–1930, Eng. transl. 1995, p. 177). "Becoming-animal" broke with essentialism, caught humans in the duration of concatenation and contagion and inaugurated "a vision of life under the sign of immanence" (Cimatti 2013, p. 130, my transl.). It enabled the recognition of hybridisations with retroactive effects on the human system, modifying, contaminating and altering its identity (Marchesini 2002, p. 115). And yet, despite the shift from a static "I" to the "process" of becoming-other – me-you or me-he/she – the subject of action and becoming is still the human being.

An alternative would be to investigate the relationship between humans and animals through the "detour-and-access" approach used by François Jullien (1995) in his study of China. Building upon Foucault's notion of heterotopian shift, Jullien suggests a delocalisation into the animal's environment and habits, that is, into its *Umwelt*. This process would be no more "animalising" that the one described by Jullien is "chinesifying": the Foucauldian heterotopian shift involves adopting the position and perspective of the animal to then return to a *cogito* that has been transformed and reconfigured by the whole process. On a metalinguistic level, if the change of becoming-animal is brought about by time (the temporal aspect) – as humans shed their own identity and *appropriate* that of the animal – the heterotopian shift is a spatial one: we *adopt* the animal's position only temporarily but by embodying its perspective. The arts and the anthropology of the ontological turn both put this shift into practice: they conceive the animal's physical nature from the animal's perspective. This enables them to verify the thesis of the embodied mind, as the physical nature of humans is not addressed directly but in negative, through the animal.

4 Animality in New Anthropology

Over the past 30 years, i.e. parallel to the development of embodiment theories, anthropology has gone through the so-called ontological turn. The pivotal studies, most of which were conducted in Latin America, called for the abandonment of ontological monism (based on the idea of a single being endowed with a single truth), for the abandonment of the idea that all entities share an identical nature and for the existence of different worldviews with different values and modes of knowledge.

	Dissimilar in Physicality	Similar in Physicality
Similar in Interiority	Animism	Totemism
Dissimilar in Interiority	Analogism	Naturalism

Fig. 1 Ontological scheme, the four modes (Descola 2005, p. 214)

In particular, Descola's (2005) classification into animistic, totemic, naturalist and analogical cosmologies, derived from Eduardo Viveiros de Castro (1998), frees us from our exclusive bond with Western naturalism, i.e. from the thesis of *the* original subject from whom everything is derived, and postulates the coexistence of multiple natures. According to Descola (*op.cit.*), animism implies inner resemblance and physical difference; totemism, inner and physical resemblance; naturalism, inner difference and physical resemblance; and analogism, inner and physical difference (Fig. 1).

4.1 Animism

What interests us in animistic mentality is the fact that it is founded on the principle of the inner continuity of the soul, hence the animistic belief in transmigration, and physical discontinuity, hence the belief that our bodies distinguish us. The soul remains formally identical throughout the species and is a reflexive form: it includes what it perceives from its own point of view; each body, to the contrary, is different and constitutes its own system of active, relational and transitive affections, expressed when becoming the object of the other's perspective. Thus, jaguars and humans are both predators from the point of view of a lamb; jaguars are predators and lambs are prey from the point of view of a human; and humans and lambs are both prey from the point of view of a jaguar (Viveiros de Castro 1998, p. 470).

This kind of "perspectivism", observed and theorised for the first time by anthropologist Tânia Stolze Lima, who studied the Juruna of Brazil, is different from relativism, which admits various points of view on a single reality (Lima 1996). There is no body in the sense of an a priori physiological essence with a series of undefined cognitions embodied in it, but a constellation of attitudes and habits that emerge in the form of points of view – that is to say, differences – and which determine the unique character of the body of a given species: what it eats and how it communicates. Knowledge is truly embodied because whatever is activated or "agented" by a point of view will be a *person*, in the sense of a concatenation of affects, affections and *habitus* – that is, as an *instance of enunciation*. Here Viveiros (1998) is amply quoting and using the linguistic and semiotic theories of Benveniste (1956, 1958).

4.1.1 Animals as Deixis

Animistic mentality raises two important reflections. The first is that the condition shared by humans and animals is not animality but humanity. The animistic model of the soul is human, while that of the body is animal. This worldview cannot imagine a polar opposition between *Homo sapiens* and animals, because it refuses any hierarchy based on the possession of language. It is no coincidence that Viveiros analysed animistic theories through the filter of Benveniste's system of the enunciation, which can also illustrate the mechanisms of extralinguistic communicative actions and reactions. The enunciation apparatus studies deictics and markers of objectivity and subjectivity in the framework of actual linguistic exchanges and speeches, i.e. *parole*, but its power of generalisation goes beyond verbal language. It lends itself quite well to describe cultures where the "subject" is not necessarily a speaking one. To the contrary, precisely because the model of the soul shared by humans and animals is the human soul, the Amerindian terms that are usually translated as "human being" do not actually refer to the natural category of the human species or to a member thereof, but to a *person*, an animated being in the pronominal sense – be it a human, a jaguar or a lamb – as a relational position (Viveiros de Castro 1998, p. 483), the marker of a presence.

Supposedly "human" characteristic are not the prerogative of humans. *They are produced within the body*. Individuals – as mentioned above – perceive the members of their own species as humanity, as culture. The species possesses a deictic *us* and, within the group, humanity is a reflexive property – the lamb is human to another lamb, and the jaguar is human to another jaguar. Contrariwise, identity, i.e. the difference among bodies that can only be perceived from an external point of view, is the product of a scopic function. The impersonal pronouns *it* and *them* indicate the non-person or the object, i.e. nature (ibidem, p. 483) Then there is the *you*, the second person (singular or plural) seen as the other subject of a nonhuman perspective, the supernatural, a dead persona or spirit. "Only shamans, multinatural beings by definition and office, are always capable of transiting the various perspectives, calling and being called 'you' by [non-human subjects] without losing their condition as human subjects" (ibidem).

In rituals, these pronominal functions are inverted. Human bodies, covered in feathers, colours, patterns, masks and other animal prostheses, are transformed into animals and reveal to their own eyes the "natural" peculiarity of their body.

Interestingly enough, the moment of the maximum social objectification of bodies coincides with the moment of maximum animalisation (ibidem, p. 480).

4.1.2 Vestments and Investments

The second issue raised by the study of animism is that the nature of things changes in function of the *habitus* of the observer and is therefore the fruit of a complex combination of watching and being watched. Just as there is no a priori subject, there is also no fixed essence of things, but only the *objective correlative* of the point of view. In particular, the nature of what is observed is at the core of a "primary dialectic […] between seeing and eating" (Mentore 1993, p. 29). "What to us is blood", writes Viveiros (1998, p. 478), "is maize beer to a jaguar; […] what we see as a muddy waterhole, the tapirs see as a great ceremonial house". What varies is "what passes through the optic nerve (or digestive tube) of each species, so to speak" (Viveiros de Castro 2012, p. 34). There are no self-identical substances that are being categorised in different ways, but relational multiplicities such as blood-beer or muddy waterhole-ceremonial house. In this dialectic:

1. Scopic functions are non-simultaneous. A human and a jaguar cannot be both "subjects", or rather "people", at the same time. A jaguar pouncing on a man sees the latter's blood as beer before the man has become aware of the peril. "Each species has to be capable of not losing sight, so to speak, of the fact that the others see themselves as people and, simultaneously, capable of forgetting this fact—that is, of—no longer seeing it" (ibidem, p. 34). This is particularly important when humans become predators, that is to say when they kill to eat. They must be the first to see so as not to be seen.
2. Although we must be capable of "not seeing" the animals we eat as they see themselves, sometimes it may be useful or even necessary to see how certain animals see and to see them as they are seen by other animals (ibidem). The self incorporates the point of view of the enemy. In order to avoid being eaten by the jaguar, we must adopt its point of view on itself, become the other by occupying the enemy's position towards us, but without letting us be overwhelmed by alterity.
3. Thus, the visible form of the body, which is a powerful sign of the differences between affections, is not a fixed entity but a variable and metamorphic appearance, an ever-changing and removable clothing. It can be deceptive: a human form may conceal a jaguar-affection. Appearances are misleading, not because they differ in their supposed essence, but precisely because they are appearances, i.e. something that appears. An appearance implies and is determined by a recipient, a subject in front of whom the act of appearing takes place. "It is not so much the body that is a clothing but rather that clothing is a body" (Viveiros de Castro 1998, p. 482).

4.2 *Animistic Mentality in the West*

The merit of Viveiros was to show that the identification modes of animism are not limited to the practices of native populations, but that they are and have always been present in our world, too. Not for religious reasons or because Buddhism is currently in fashion. Animistic logic, whereby body and soul are effects of perspective and never a priori categories, may explain wartime camouflage, the mimicry and animal-like movements of soldiers, animal-rights movements as well as the recent evolution of eating habits and therefore of the food industry (Marrone 2011). Let us think of the controversy about eating meat. The main charge brought forward by vegans is that, by ingesting the flesh of animals, meat eaters also ingest the soul that these animals have in common with us. The argument of the defence is that, deep down, we are all cannibals, as humans always eat an alterity which they reduce to themselves through a variety of beliefs and rituals (Lévi-Strauss 2013). The former argument is "driven" by the firm *animistic* belief that animals embody a soul that we share, i.e. that their body, being "an assemblage of affects or ways of being that constitute a *habitus*" (Viveiros de Castro 1998, p. 478), incorporates a reflexive form that is equivalent to ours; the second argument, which is *analogic*, is ruled by a basic "exterior"/"interior" dichotomy between humans and the animals that humans reduce to their own life form. Only "a theory of the sign and communication" (Viveiros de Castro 2012, p. 35), i.e. semiotics, can give an explicit form to these controversies.

As proof that animism, just like any other worldview, has no geographic location but is a question of thought and beliefs, Viveiros borrows not only the notion of enunciation from Benveniste but also the idea of the soul as a concatenation of affects, affections and *habitus* from Deleuze: here "affect" is the acting power that increases or decreases according to the chance encounters in which the sentient body takes part; and "affection" is a composition of bodies, a relationship of mutual capture that triggers the alterations imposed by a body to another (Deleuze 1980). Deleuze, despite being a Western philosopher, is not a naturalist. His reference is another Western philosopher, Spinoza, who defined the affectus as "the modifications of the body, whereby the active power of the said body is increased or diminished, aided or constrained, and also the ideas of such modifications" (Spinoza 1677, Eng. trans. 2001, p. 4). Ethology itself, according to Deleuze, is based on the affects of which we are capable and which define the bodies:

> The approach is no less valid for us, for human beings, than for animals, because no one knows ahead of time the affects one is capable of; it is a long affair of experimentation, requiring a lasting prudence, a Spinozan wisdom that implies the construction of a plane of immanence or consistency. Spinoza's ethics has nothing to do with orality; he conceives it as an ethology, that is, a composition of fast and slow speeds, of capacities for affecting and being affected on this plane of immanence. (Deleuze 1970, p. 125)

The interpretations of perspectivism in Lima, Viveiros and later Descola, also draw on Deleuze's "truth of relativity" as opposed to the "relativity of what is true": in a non-essentialist and phenomenical worldview, the subject is an event: a vector

Fig. 2 Photogram from the American computer-animated comedy film *Madagascar 1* (2005), DreamWorks Animation

of curvature, the effect of what occupies the point of view and makes it what it is (Deleuze 1988), "pure powers whose act consists in habitus or arrangements (folds) in the soul" (ibid., Eng. trans. 2006: 24).[2]

Far from lending itself to criticism, the use of Western categories to explain animism merely demonstrates the existence of approaches alternative to naturalism in the West. Which is why we are not surprised by a scene such as the one featured in the animation film *Madagascar 1* (2005), where Alex, a lion crazed by hunger, sees his zebra friend Marty as a steak (Figs. 2 and 3).

We may not be aware that what is at play is the change of affection of an animist, but the meaning of what is going on is perfectly clear. The four worldviews coexist – indeed, who is more totemic than a fetishist? – and do not belong to different worlds. In *Madagascar 1*, the commutation between zebra and steak illustrates the conception of things and subjects as relational terms: physical entities "vested" with the investments of the beholder and expressing and translating the latter's affections and affects. In this specific case, the contrast between the zebra's envelope-container and the meat-content expresses, semi-symbolically, its transformation from subject-entity to object-prey:

[2] On perspectivism in the anthropology of the ontological turn, see Migliore (2015).

Fig. 3 Photogram from the American computer-animated comedy film *Madagascar 1* (2005), DreamWorks Animation

Fig. 4 Vegan sensibilisation campaign against the consumption of animal meat www.wired.it/lifestyle/food

Black-and-white striped exterior: blood-red interior:: subject-entity: object-prey

Many pro-vegan campaigns resort to a similar strategy: they intimidate meat eaters by imposing their own animistic worldview (Fig. 4).

The anthropology of the ontological turn, which provides the tools to reflect upon these differences, brings up the proximal rather than distal relationship between animals and humans. The category /human/ cannot be defined without the category /animal/: in other words, /human/ and /animal/ mutually define each other. Moreover, the very notions of survival and social objectification, as well as civilisation, require the capacity to incorporate the animal's perspective. Thus, animals are not "nonhuman": "human" is a complex category" that includes the participatory subcategory/human/-/animal/. Viveiros de Castro (1998, p. 472) goes so far as to say that *Animals are ex-humans (rather than humans, ex-animals)*. "The myths tell how animals lost the qualities inherited or retained by humans. Humans are those who continue as they have always been". "While our folk anthropology holds that humans have an original animal nature that must be coped with by culture – having been wholly animals, we remain animals 'at bottom' – Amerindian thought holds that, having been human, animals must still be human, albeit in an unapparent way. Thus, many animal species, as well as sundry other types of nonhuman beings, are supposed to have a spiritual component that qualifies them as 'people'".

5 Animality in the Arts

Visual arts started with the creation of images that illustrated the continuity and contiguity between humans and animals. The first prehistoric pictorial subjects were animals painted with animal blood: these bison were not meant to represent a given event but rather to anticipate and bring luck to future hunting parties. The subjects, painted with the blood of preys, expressed the sacred nature of the animal that would save the group from starving (thanks to its meat) or freezing to death (thanks to its skin and furs).

> An existential relationship was established between humans and animals. Animal life nourished human life [...]. The meal was the act through which spiritual symbiosis was achieved and completed by means of the integration of the animal spirit into a human body [...]. The representations of anthropomorphic beings masked as animals and of anthropo-zoomorphic beings in hunting art are a spectacular illustration of this search for symbiosis [...]. The death of one gives life to the other. The dead animal was a source of food and life. Through the physical assimilation of the flesh of the animal, humans also acquired its strength, vitality and all its capacities, both real and imaginary. (Anati 2002, p. 87, my translation)

Animal parts, secretions and other remains soon became ingredients in the preparation of colours (Elkins 1999), contributing to the painting's pictorial texture. Even the glues used in Renaissance painting were made from horse hooves, deer antlers or rabbit skins. Some glues used in modern painting are still made with animal skins. On the figurative front, the preservation of animal traits in the artwork was conveyed through two main channels: on the one hand, the animal vital force found its iconographic expression in the miniatures and on the capitals and gargoyles of cathedrals, which made reference to Medieval bestiaries; on the other, this energy symbolically functions as a metaphor for the survival of the artwork itself – a metaphor ironically overturned in the interest in taxidermy and formaldehyde manifested by contemporary artists such as Annette Messager, Damien Hirst, or

Fig. 5 Georg Baselitz, *Portrait und Adler*, 1978, 250 × 340 cm (Private Collection)

Maurizio Cattelan. We should also mention the huge iconological contribution to zoosemiotic imagery and to the linguistic and metalinguistic meanings of animals, from lions to parrots and monkeys. Nowadays, numerous artists disguise themselves as animals (Ana Mendieta, Joan Jonas, Jan Fabre) or represent what they observe *sub specie animalis* (Alberto Savinio, Jean Dubuffet, Joseph Beuys).

Several impressive experiments have been carried out in order to develop an *animal phenomenology* (Broglio 2011), i.e. to work on the notion of interspecies through a heterotopian shift into the animal world (see Ramos ed., 2016). A case in point is Georg Baselitz's *Portrait und Adler* (1978), a 250 × 340 cm diptych painted with oil and tempera on plywood (Fig. 5).

The diptych features two upside-down figures, painted in the same blue tones, with the same red shadows, black-and-white outlines and grey background: the full-body figure of an eagle, its beak turned to the side, on the left panel; and the head and shoulders of a man, seen from the front, on the right one.

Although the two backgrounds are painted in the same colour, the tones on the left are bright and saturated while those on the right are more blurred, suggesting that the two portraits may have been painted in the same place but at different times, dynamically and on the spot. The image freeze captures the two figures in what appears to be a reverse chronological order with respect to the title: following the Western left-to-right reading direction, the man (*Portrait*) should be on the left and the eagle (*Adler*) on the right. This is not the case, and only an aerial perspective from the bottom could justify the inversion and upside-down position. It is as if the perspective were that of a bird, perhaps another eagle, seeing an animal of its own species (a similar *zoon*) to its left (*Adler*) and an animal from a different species, a (*zoon politikon*) to its right (*Portrait*). What remains is the verbo-visual contradiction: *Adler*, in German, is both the common noun of the eagle and a proper noun; moreover, due to its character, the eagle has been the symbol of Germany ever since the times of Scandinavian myths – undergoing a series of semantisations and polarisations throughout the centuries.

Which figure, then, designates the *Portrait* and which one the *Adler* in Baselitz's painting? The answer lies in the animal staging of a mutual capture.

References

Anati, E. (2002). *La struttura elementare dell'arte*. Capo di Ponte: Edizioni del Centro.

Baldinucci, F. (1681). *Vocabolario Toscano dell'Arte del Disegno*. New Edition: Baldinucci, F. (1809). *Società Tipografica de'Classici Italiani, Milano*

Belpoliti, M. (2015). *La strategia della farfalla*. Parma: Guanda.

Benveniste, E. (1956). La nature des pronoms. In: M. Halle & R. Jakobson (Eds.), *For Roman Jakobson* (pp. 34–37). The Hague: Mouton & Co. Italian Edition: Benveniste, E. (2009). La natura dei pronomi. In P. Fabbri (Ed.), *Essere di parola. Semantica, soggettività, cultura* (pp. 138–143). Milano: Bruno Mondadori.

Benveniste, E. (1958). De la subjectivité dans le langage. *Journal de Psychologie, 55*, 257–265. Italian Edition: Benveniste, E. (2009). Della soggettività nel linguaggio. In P. Fabbri (Ed.), *Essere di parola. Semantica, soggettività, cultura* (pp. 111–118). Milano: Bruno Mondadori.

Broglio, R. (2011). *Surface encounters: Thinking with animals and art*. Minneapolis and London: University of Minnesota Press.

Caillois, R. (1960). *Méduse et Cie*. Paris: Gallimard. English Edition: Caillois, R. (1964). *The mask of Medusa: the relationship and contrast between insect and man*. New York: Clarkson N. Potter.

Casarin, C., & Fornari, D. (2010). *Estetiche del camouflage*. Et al, Milano

Cimatti, F. (2013). *Filosofia dell'animalità*. Roma-Bari: Laterza.

Clark, A. (2008). *Supersizing the mind: Embodiment, action, and cognitive extension*. Oxford: Oxford University Press.

Clark, A., & Chalmers, D. (1998). The extended mind. *Analysis, 58*, 7–19.

Deleuze, G. (1970). *Spinoza: Philosophie pratique*. Paris: PUF. English Edition: Deleuze, G. (1988). *Practical philosophy*. San Francisco: City Light Books.

Deleuze, G. (1980). De quoi une corps est-il capable? Cours sur Spinoza. Available http://www2.univ-paris8.fr/deleuze/rubrique.php3?id_rubrique=6. Accessed 6 Jul 2017. Italian Edition: Deleuze, G. (2013). Cosa può un corpo? Lezioni su Spinoza. Ombre Corte, Verona

Deleuze, G., & Guattari, F. (1980). *Mille plateaux: Capitalisme et schizophrénie*. Paris: Seuil. English Edition: Deleuze, G., & Guattari, F. (2003). *A thousand plateaus: Capitalism and schizophrenia*. Minneapolis: University of Minnesota Press.

Descola, Ph. (2005). *Par-delà nature et culture*. Paris: Gallimard. English Edition: Descola, Ph. (2013). *Beyond nature and culture*. Chicago: University of Chicago Press.

Eco, U. (2006). *A passo di gambero. Guerre calde e populismo mediatico*. Milano: Bompiani.

Elkins, J. (1999). *What painting is. How to think about oil painting, using the language of alchemy*. New York/London: Routledge. Italian Edition: Elkins, J. (2012). *La pittura cos'è. Un linguaggio alchemico*. Milano: Mimesis.

Fabbri, P. (1991). *Il significante del mondo*. New Edition In: Fabbri, P., & Marrone, G. (Eds.) (2001). *Semiotica in nuce II* (pp. 328–335). Roma: Meltemi.

Fabbri, P. (2008). Estrategias del camuflaje – Entrevista de T Migliore. *Revista de Occidente*. José Ortega y Gasset, Madrid, Noviembre, 330, 89–110.

Fabbri, P. (2011). Semiotica e camouflage. In S. Scalabroni (Ed.), *Falso e falsi. Prospettive teoriche e proposte di analisi* (pp. 11–25). Pisa: ETS.

Fontanille, J. (2004). *Soma et séma. Figures du corps*. Paris: Maisonneuve et Larose. Italian Edition: Fontanille, J. (2004). *Figure del corpo. Per una semiotica dell'impronta*. Roma: Meltemi.

Gombrich, E., & Kris, E. (1938). The principle of caricature. *British Journal of Medical Psychology, XVII*, 319–342.

Heidegger, M. (1929–1930). *Grundbegriffe der Metaphysik*. English Edition: Heidegger, M. (1995). *The fundamental concepts of metaphysics. World, Finitude, Solitude*. Bloomington: Indiana University Press.

Jullien, F. (1995). *Le détour et l'accès: Stratégies du sens en Chine, en Grèce*. Paris: Grasset. English Edition: Jullien, F. (2000). *Detour and access, strategies of meaning in China and Greece*. New York: Zone Books.

Lakoff, G., & Johnson, M. (1999). *Philosophy in the flesh: The embodied mind and its challenge to western thought*. New York: Basic Books.

Lévi-Strauss, C. (2013). *Nous sommes tous des cannibales*. Paris: Seuil. English Edition: Lévi-Strauss, C. (2015). *We are all cannibals: And other essays*. New York: Columbia University Press.

Lima, T. S. (1996). O dois e seu múltiplo: reflexões sobre o perspectivismo em uma cosmologia tupi. *Mana, 2*(2), 21–47. English Edition: Lima, T. S. (1999). The two and its many: Reflections on perspectivism in a Tupi cosmology. *Ethnos, 64*(1), 107–131.

Marchesini, R. (2002). *Post-human. Verso nuovi modelli di esistenza*. Torino: Bollati Boringhieri.

Marrone, G. (2005). *La Cura Ludovico. Sofferenze e beatitudini di un corpo sociale*. Torino: Einaudi.

Marrone, G. (2011). *Addio alla natura*. Torino: Einaudi.

Mentore, G. (1993). Tempering the social self: Body adornment, vital substance, and knowledge among the Waiwai. *Journal of Archaeology and Anthropology, 9*, 22–34.

Merleau-Ponty, M. (1964). *Le Visible et l'Invisible suivi de Notes de travail*. Paris: Gallimard. English Edition: Merleau-Ponty, M (1968). *The visible and the invisible*. Evanston: Northwestern University Press.

Migliore, T. (2008). L'efficacia del segreto. Strategie del mimetismo tra natura, guerra e arti. In T. Migliore (Ed.), *Argomentare il visibile. Esercizi di retorica dell'immagine* (pp. 223–231). Bologna: Esculapio.

Migliore, T. (2015). Sul prospettivismo. In G. Ferraro, A. Giannitrapani, G. Marrone, & S. Traini (Eds.), *Dire la natura. Ambiente e significazione* (pp. 205–221). Roma: Aracne.

Ramos, F. (Ed.). (2016). *Animals*. Cambridge, MA: MIT Press.

Spinoza, B. (1677) *Ethica ordine geometrico demonstrata. In: Opera posthuma*. Amsterdam: Jan Rieuwertsz. English Edition: Spinoza, B. (1951). Ethics, demonstrated in geometrical order. In: R. H. M. Elwes (Ed.), *The chief works of Benedict De Spinoza*. New York: Dover.

Stengers, I. (1997). *Cosmopolitiques, I, II*. Paris: La Découverte. English Edition: Stengers, I. (2010). *Cosmopolitics I, II*. Minneapolis: University of Minnesota Press.

Thom, R. (1988). *Esquisse d'une sémiophysique*. Paris: InterEditions.

Varela, F., Thompson, E., & Rosch, E. (1991). *The embodied mind. Cognitive science and human experience*. Cambridge, MA: MIT Press.

Viveiros de Castro, E. (1998). Cosmological deixis and Amerindian perspectivism. *The Journal of the Royal Anthropological Institute, 4*(3), 469–488.

Viveiros de Castro, E. (2012). Immanence and fear: Stranger-events and subjects in Amazonia. *HAU: Journal of Ethnographic Theory, 2*(1), 27–43.

Wilson, R. (2004). *Boundaries of the mind: The individual in the fragile sciences*. Cambridge: Cambridge University Press.

The "Morally Abandoned Child" and the "Inner Savage"

Pierluigi Cervelli

Abstract "The day of February 13th, 1921, a barefoot little girl came out of one of Trastevere's alleys and walked slowly along King's Avenue. The little girl has a pale, sad face, she has unkempt hair, she's dirty and laid bare feet on the last fallen leaves so she did not feel the cold off the soil." This short text constitutes the opening words of a book, published in 1926 in Rome, which Maria Capozzi – director of a center for "morally abandoned children" – publishes with the support of the governor (the governor is the government-appointed institution (nominated directly by Mussolini) replacing the mayor and the city council elected democratically in the city of Rome) of the city to document the activities of the "Infantiae Salus institutes." The hypothesis underlying the interest for this text is that following the great upheaval of the First World War, with the advent of fascism begins a profound process of modeling Italian identity.

> The day of February 13th, 1921, a barefoot little girl came out of one of Trastevere's alleys and walked slowly along King's Avenue. The little girl has a pale, sad face, she has unkempt hair, she's dirty and laid bare feet on the last fallen leaves so she did not feel the cold off the soil.
>
> The Avenue is almost deserted: an hour before so many chatting kids, each with a bag and a little basket, passed by on their way to school: many men, and many women passed, all headed for work, but now people are fewer and rushing. Only the little vagabond, who has left the house without a purpose, will wander all day aimlessly, begging alms to passers-by as she does every day, it is just that the little one walks slowly, skimming through the trees uncertain, and the cold wind of that morning without sunshine makes her all shiver.
>
> Once at the Girolamo Induno Street, she randomly turns at the corner, perhaps because she feels that the wind from that side blows less, and then follows a rustic wall that borders the road.
>
> At some point she stops surprised: voices of children singing and laughing reach from behind the wall. The little girl makes yet a few more steps and reaches a gate, but an iron plate on the bars prevents her from seeing: so she draws the face between the wall and the gate near the hinges where there is a small empty space, and looks.
>
> In front of her absorbed look, chasing children pass over and over again, as they play cheerfully, content and with ruddy faces, and they seem to not mind the bitter cold. They all

P. Cervelli (✉)
University of Rome, La Sapienza, Italy
e-mail: pierluigi.cervelli@uniroma1.it

G. Marrone, D. Mangano (eds.), *Semiotics of Animals in Culture*, Biosemiotics 17, https://doi.org/10.1007/978-3-319-72992-3_13

have a white apron, and over their apron a jacket and a scarf that keeps them warm: they are all clean, well dressed with shoes and are happy and smiling. The little observer also sees well-ordered flowerbeds, pots with plants: a very feast of purity and green.

For how long does she remain full of admiration, taken by the wave of cheerfulness and well-being that keeps her there still?

But here a child spots the tiny attentive face outside the gate, approaches her quickly and puts his face against hers: – Who are you? – She shakes and pulls back. Other kids come closer, look at her, and talk to her: – Who are you? What do you want? What do you do? All of a sudden the little curious ones disappear at the sound of a voice: the gate squeaks, and as it opens a young lady appears, who having glanced for an instance at the poor numb little girl, extends her hand with an endearing smile.

"Little one, what are you doing? Do you want something?" – The little one does not answer, she would have liked to run away; but all those attentive eyes that observe her curiously, keep her there nailed and withdrawn.

"Do you want to come in? Come on?" – A warm hand gets hold of her freezing hand, doing sweet violence to her. She resists, does not want to come in, she is intimidated, but hears being called by name by one of the children, and she recognizes, in astonishment, a kid who lives nearby her; she then surrenders to the hand that is pulling her. The gate closes behind her shoulders and she finds herself in a beautiful garden, surrounded by many children, among whom she recognises a few more.

After a while, the young lady lines up the children in a row and makes them go back into the building which limits the garden, and then takes the little guest by the hand, and hands her over to another young lady, who takes her over and lovingly quizzes her and encourages her.

After an hour the little girl enters the classroom where all the students are gathered: she also wears a white apron and is clean and well-combed; her class-mates find it hard to recognize her. She sits in the place assigned to her, and tells her name to the teacher: Panella Giuseppina. From that moment her life has changed: a new life unfolds before her.

This has really happened. There is on Girolamo Induno Street, an Infantiae Salus institute, where the abandoned children are welcomed and where in February of 1921 Panella Giuseppina, frigid-cold, barefoot, almost naked, was welcomed in the way I said. (Capozzi 1926, pp. 7–9)

This short text constitutes the opening words of a book, published in 1926 in Rome, which Maria Capozzi – director of a center for "morally abandoned children" – publishes with the support of the governatorate[1] of the city to document the activities of the "Infantiae Salus institutes." The preamble from lawyer Raffaello Ricci, a delegate of the governorate of Rome for childcare services, immediately clarifies the public, officials, and in some way the politics of the text. The hypothesis underlying the interest for this text is that following the great upheaval of the First World War – a real cultural explosion in the sense indicated by J.M. Lotman (1993) – with the advent of fascism begins a profound process of modeling Italian identity (that Lotman called "meta-description").

This process is structured by giving a narrative form to the new hybrid figures – outside the social order preceding the war that populates now the city (poor families, veterans, displaced persons, or immigrants, from the countryside surrounding of Rome, who once in the city immediately start living in slums).

[1]The governatorate is the government-appointed institution (nominated directly by Mussolini) replacing the mayor and the city council elected democratically in the city of Rome.

If, as Lotman has argued, in the moments of cultural explosion, the causal and scientific explanations are replaced by "explosions of metaphors," then the text we propose to analyze, we believe, represents an additional attempt to produce intelligence of these phenomena by finding a place in the semantic universe of the time, to these figures that just appeared after the First World War through their inclusion in a series of narratives.

It is therefore about giving a narrative organization to "metaphorical explosions" so as to transform them into powerful narratives, capable of redefining the relationships between believing and knowing (see Greimas 1983). Regardless of its correspondence with a factual reality, which I do not consider to be the primary object of a semiotic consideration, I find of interest the effectiveness that was to be attributed to a tale of this kind. I believe that this indicates – if not a truth – at least the uncertain limits of a *plausibility*, a verisimilitude that shaped the common sense at the time related to children and signaled, perhaps more in depth, the initial transformation of the categories at the basis of a certain semantic universe.

The narrative micro-processes that give shape to this new "master narrative" of Italian society unravel in the pages of the magazine of the governorate of Rome, *Capitolium*. Some of the border figures belong to this one, those of the dangerous subject – or at least disturbing – found in the political and educational discourse at the beginning of Italian fascism, particularly between 1925 and 1935.

Among these, one notices the appearance of a strange figure, which, in Foucault's[2] words, could be defined as an "inner savage": a presumably pre-cultural and pre-social being, borderline between the human world and the animal world. This is the "morally abandoned child," what we could call the "street child" today, one of the new social/narrative actors, which emerges most frequently in the discourse of educators and politicians.[3] I believe that for this reason, it also allows us to reflect, within the general semantic field of animality, on the limited portion of meaning occupied by what could be termed "bestial," that is, the bordering territory between human and the animal.

They are "the abandoned children who spend all day on the street unguarded [...] abandoned children who have no home and cannot attend school" (Capozzi 1926, p. 9). Always Maria Capozzi describes the origin of the phenomenon:

---

[2] In his reflection on the appearance of state racism (1976, 45), M. Foucault introduces the concept of "internal racism," which is particularly useful here: it is a racism that some societies do not address versus the exotic space and potentially undefined of the territories to colonize, but toward their inner selves, by identifying some layers of them as places to "colonize" or to identify as the bottom layers of the social hierarchy. It seems to us that what happens in the case of the text that we propose to describe, that we seem to be looking for the identification of the social subject of the unworthiness (for the suggestion of the subject of the unworthiness we thank F. Marsciani, private conversation).

[3] I draw this information through the banner of the *Capitolium* magazine, the official body of the governorate of Rome, which is presented as a "review of the governorate's activities." Publications start/date back to 1925. For more information on child-control activities, Cervelli, P., in process of publication.

> It is also noted that the plague of promiscuity is almost generalized. From such promiscuity a state of immorality is created, that contaminates and destroys families and transforms these innocents into already perverse beings who know everything, who see everything and talk maliciously of the most obscure things and naturally imitate the most shameful acts. (Capozzi 1926, p. 19)

Let us now shift to the analysis of the text within the structural semiotic theoretical framework. First of all, we are dealing with a text that is presented as true yet entirely structured as a fairy tale. It is the story of a regeneration: the inaugural, almost mythical scene, of a cultural rebirth and of a re-enrollment in society.

Despite it having the narrative structure of a fairy tale, the tale is presented as unquestionably true as it begins with the indication of a precise unit of time and space – and therefore through a speech act aiming at an objectivity effect: it is 13/2/1921 and we are in Trastevere, a poor and sickly Roman neighborhood.

Obviously, in the sense that it unfolds an effect of reality, it is also the closure of the text, ending with the indication of the name of the protagonist, presented at the beginning as an emblematic and tragic figure, typical of a saga.

Despite this discursive organization, the tale in its entirety has an elementary narrative structure of the fairy tales analyzed by Vladimir Propp. It tells the transformation – almost magical, so abrupt and simple – of one subject into another: the transition from the miserable daily life of a "little vagabond" into a different little girl, whose "new life" unfolds before her.

Underneath the motifs and the surface figures unfolds the story of an initial lack of moral compensated by a manipulating sender actant[4] that through a "sweet violence" submits to the deeds of competent subjects the small body to be transformed. By "culturalizing" it, we could say, through a semiotic organization of the body (the combing) and of the dressing (the shirts), able to make her "civil", so that she is identical to others: to transform – in other words – the little girl into the elementary member of a homogenous society, whose belonging to it – the new life, marked by the almost lack of recognition of other children – is sanctioned by the return, through its own name, of an individual and social status.

What are the traits of this "culturalization" of the little girl? And which ones of the deconstructing "naturalization" that have characterized the first appearance?

At the beginning of the story, the child is dirty, barefoot, unkempt, and above all alone: she leaves the house alone and prepares to survive by herself by means of begging. She is without a purpose, wandering randomly along a large empty boulevard – an open and indefinite space – where society, exemplified by schoolchildren and workers already running about their business, has already disappeared. Suddenly she discovers a show that captures her. A closed space marked by joy ("cheerfulness"), order ("purity"), and health ("well-being"), which is the tenet of the primary binary opposition that governs the whole text: this is obviously about her world's meaning being overthrown.

---

[4] Following the A.J. Greimas model of narrative organization, an actant is an abstract subject of action. The sender is the one that guarantees the fundamental values of the narrative.

In the safety of a closed garden – and not in the open void where she moves – the text describes in great detail, making them almost visible, children marked by gestures of care, i.e., subjected to the action of a competent person who protects them (the clothes they wear – named one by one, "sweetened" by the endearing words and described in detail – can keep them away from the cold, while she is "barefoot" and "poorly dressed"), they are "clean" while she is "dirty," and, at the end, they play, instead of working, as she would have had to do (they live their childhood time properly).

Irresistibly confined to the glimpse,[5] the little girl is noticed, and the meeting with the other children has the taste of a story of first contact with the radical "other" but by the overturned path: it is in fact she who occupies the position of the other who looks stunned, almost thunderstruck. Called several times, in fact, the "little observer", struck, perhaps shaken, she does not respond. And here appears a reassuring voice of authority, the competent subject who will be the operator of the transformation: "a young lady" – an actor surrounded by figures of calmness, safety, and knowledge – that will make this "intrigued poor little girl" to become the pupil of a school subjecting her to "sweet violence."

Once inside the garden, where she discovers that the small protagonists of that strange world are in fact familiar to her ("recognizes many others"), the little girl enters an entity that provides children with a material form, in the collective space of the school, and in the inside of the individual body: another schoolteacher will comb her properly, wash her, dress her a white apron, assign her a specific place in the inside of a whole, and in return ask her "only" to contribute to the formation of a knowledge: "she questions her and encourages her."

In only 1 h, the miracle is accomplished: "The class-mates find it hard to recognize her." What used to be a "messy" and "intrigued" little girl has assumed the same brilliance and the same order of space as the one she is fitted into: her body is perfectly rhymed – that is to say, we can say it is characterized by a categorical isomorphism – with the collective space where she is integrated.

It is worth noting that the little girl remains completely passive, and in some way she even resists to the good that is offered to her: she would like to run away, but, struck by the charm that the place emanates, she remains "nailed and withdrawn," and in the end, just because of a child who already knows her name, she concedes. The competence to act is therefore entirely on the side of the institution: the little girl, described at the beginning of the tale as discarded waste – the amorphous social matter constituted by a small vagabond with whom no one deals with – is the object,[6]

[5] The glance in this case between the little girl and the schoolteacher, or the other children, recalls the immobility and the valorization of the distance – without words – typical of the sudden glimpse exchanged between the human being and the wild animal met by chance/of which Francesco Marsciani spoke at the conference in which a first version of the essay was presented as a report/paper.

[6] It is interesting to note that in the theoretical framework outlined by E. Landowski (2005), the little girl could be considered the subject of a semantic programming regime aimed at producing "normal" individuals that represent the reproduction of an ideal social type. What is important is that, as Landowski notes, within such a way of thinking, a human subject becomes an object without will.

throughout the narration, of an effective manipulative transformation that will eventually make her assume, for the first time throughout the text and, as narration suggests, also in life, an identity (she is no longer wandering out of time and without a place, but has a precise place in an organized system). Having become a schoolgirl, she can finally pronounce her name.

As we see in the narration, no explicit animal figures or direct animal-like comparisons appear. The bond is implied; the description implies a process of naturalization, so the little girl appears – because of the abandonment and the solitude – led by primary instincts: avoiding the cold and satisfying hunger (through alms).

Perhaps that is why the story recalls very closely the semantization of the wild recognizable in Darwin's diary[7] and thoroughly analyzed by Lancioni (2009), whose traits are all present in the little girl's description: it is the non-separation from "nature" due to the "dressing" semiotics ("barefoot and poorly dressed"), the de-construction of the social system (no one deals with the little girl and she does not have a real family; even if she is not abandoned, it is as if she were, hence the category of moral abandonment), and finally the natural mimetic ability, as already noted by Maria Capozzi, which is like that of the wild Fuegian of which Darwin speaks, an unconscious imitation.

Finally, common to this series of differences is the repetition of the modal opposition between constraint (unable to do and not be able to do it) typical of the Fuegian and the little girl (which manifests itself in the impossibility of choosing: the little girl will wander without a purpose, will beg as she does every day, and will not be able to decide whether or not to go in the school) that is opposed to the freedom of the children in school (freedom of play, of the proper relationship with adults, to live as children). What is also interesting is, however, that the situation is not presented as a condition of original lack of culture, but as a "return to nature," which paradoxically has nothing natural, given that it is constructed for denial or for opposition compared to the positively valorized "cultural" practices: in the very body of the little girl, as in the case analyzed by Marrone (2005), "there is no spontaneity or naturalness," and even hers is already, from its very beginning, in its very own bestiality and yet prior to the treatment – a "social body."

References

Capozzi, M. (1926). *I padiglioni Infantiæ Salus*. Roma: Unioni arti grafiche abruzzesi.

Greimas, A. J. (1983). *Du sens II*. Paris: Seuil.

Landowski, E. (2005). *Les interactions risquées* (Actes Sémiotiques) (pp. 101–103). Limoges: Presses universitaires de Limoges

Lancioni, T. (2009). *Immagini narrate Mondadori Università*. Milano.

Lotman, J. M. (1993). *Cercare la strada*. Venezia: Cortina.

Marrone, G. (2005). *La cura Ludovico Einaudi, Torino,* English Edn. Toronto: Ludovico's Cure, Legas.

[7] I refer to Darwin diary "The voyage of the beagle", analyzed by Lancioni (2009).

Frank and Johnny with Evie: Ontological Shifts in a J.R. Ackerley Novel

Francesco Mangiapane

Abstract This contribution tries to investigate the question of how to frame, from the point of view of the affiliation models, the relationship with pets. In this regard, *We Think the World of You* reveals as an exemplary text. The novel in question will be considered as a privileged place where the experience of relationship with animals is assessed, while programs and forms of an interspecies living together take form.

1 Preamble

This article proposes the analysis of *We Think the World of You*, an autobiographical novel by Joe Ackerley, published in London in 1960 with the purpose of questioning it as an exemplary text for the semiotic description of the relationship between humans and pets. Rather than reconstructing an alleged "animal language" independent from the relationship with manhood, the aim of this article is to "cut through the field"[1] of a Zoosemiotics 2.0 able to put the focus of its investigation on the cultural issue that sees men and animals, humans and nonhumans, interacting in social groups and communities. It is, therefore, with this goal, that we propose a semiotic reinterpretation of a little classic of literature about animals, believing that the literary text can be considered as an exploration of the forms of this same coexistence. The novel in question will be considered as a privileged place where the experience of relationship with animals is assessed, while programs and forms of an interspecies living together take form.

We chose to work on the relationship between humans and dogs, by virtue of their old friendship, while caring of selecting a text as far as possible away from the stereotypes that characterize the stories about pets. In short, it was preferable to temporarily exclude – but it might be well worth getting back to this ever else – those stories that mechanically associate animal species and cardinal virtues (Lassie's fidelity, Rin Tin Tin's courage, etc.). The perspective here chosen is

[1]Following the invitation that Paolo Fabbri addressed to the audience of the conference *Zoosemiotica 2.0 Forme e Politiche dell'animalità*, held in Palermo, 1–2 December 2016

F. Mangiapane (✉)
University of Palermo, Palermo, Italy

G. Marrone, D. Mangano (eds.), *Semiotics of Animals in Culture*,
Biosemiotics 17, https://doi.org/10.1007/978-3-319-72992-3_14

interested on those narratives which tackle the difference of species, staging characters who are questioned on the meaning of their experience of relationship with the alterity represented by their pet. Moreover, the tale chosen is focused on a puppy. The very notion of "puppy" combines humans and animals within a further relevance that puts the difference of species in brackets in the name of a wider and deeper identity of living beings. Focusing on the role that puppies play in the specific narrative in question will thus be a way of highlighting the affinity between subjectivities commonly thought of as irreducible. We will consider puppies as prime wrecker of a constituted order which claims men and animals as entities belonging to separate worlds. The practice of relationship that novels like *We Think the World of You* stress will thus represent a way to denounce – if ever needed – the arbitrariness of such a border and, secondly, a way of being led to the new shapes that it may take throughout the tale.

2 Notes on the Novel

We Think the World of You is the second (1960) of two autobiographical novels – the first being entitled *My Dog Tulip* (1956) – that Joe Ackerley, a writer and journalist, traveler and television man, devotes to the relationship with his dog. Both have, over time, become true classics of literature about animals, beloved by readers of all over the world, thanks also to the success of their cinematic transpositions.[2] The two novels are strongly related with each other, telling the story of the same dog in two different moments of its life: *We Think the World of You* follows its fortunes from the first encounter with its master until maturity, stage which is investigated in *My Dog Tulip*.

Although they are centered on the same canine actor,[3] the novels are very different in style and tone. *We Think the World of You* tells of the "liberation struggle" that the protagonist hires to knock out the Alsace shepherd puppy who he falls in love with from the dense plot of affections in which it is harnessed. *My Dog Tulip*, on the other hand, focuses on describing the exclusive, absolute, and poetic relationship between the master and his dog, now redeemed and free from any legacy with its previous life.

Curiously, the two novels are published in a reverse order, so that *We Think the World of You* ends up representing a sort of *Tulip*'s prequel, as if to emphasize that every true love between men and animals, the kind of the one described *My Dog Tulip*, can only be founded (*background*) on emancipation and mutual recognition. *We Think the World of You* is thus called to account for such an aspiration to self-determination which, predictably, will not only affect the dog but also its master.

[2] The first film, *We Think the World of You*, 1988, directed by Colin Gregg with Alan Bates and Gary Oldman. The second, *My Dog Tulip*, is a 2009 animated feature directed by Paul and Sandra Fierlinger, with dubbing of actors known as Christopher Plummer, Lynn Redgrave, and Isabella Rossellini.

[3] It's the she-puppy Queenie renamed Evie in *We Think the World of You* and Tulip in *My Dog Tulip*.

We like to launch the idea that both novels represent a literary attempt to address the ambiguity, at the same time semantic and identity-related, in which interspecies relationships, between human and animal beings, live. The two novels propose an exploration around the profound meaning of relating with *pets*, questioning their character and their place in the world. The inquiry proposed by Ackerley, articulated through the two novels, leads the protagonist to acknowledge the ineffability of the man-animal boundary, recognizing the terms of a brand new identity through the relationship with his pet. The way such awareness is earned is well worth being analyzed.

This work, however, focuses exclusively on *We Think the World of You*. In this text, the meaning of the life of the she-dog *Evie* emerges by difference with that of a third subject, Dickie, another puppy – this one, human – to which the dog is constantly compared by all the characters in the tale. The underlying hypothesis of the analysis here presented is that, in the "liberation struggle" hired by the master to redeem the destiny of his puppy of choice, it, as a pet, will act as a "blank sign." This "emptiness" is just waiting to be filled with affections, ties, and heterodox statuses: *Evie* ends up turning into a sort of universal signifier, the perfect recipient of the goodness of the world.

2.1 Plot

In mid-twentieth England, Frank and Johnny are lovers. Their relationship is marked by the class difference. Frank, an intellectual and libertine, is in love with his hustler, gathered among the poor people of the London suburbs. Frank appears protective and indulgent toward his boyfriend, despite the fact that he duly ends up getting into troubles. Besides Johnny's arrest for attempting to ransom an apartment, many issues need to be faced: how to arrange, during the upcoming detention, his children and pregnant wife, how to deal with the money that are never enough, and – small detail – who to entrust Evie to, the Alsatian shepherd's puppy whose Johnny declares to "think the world."

Johnny turns to Frank, asking him to take care of his pet, but he refuses: he does not know anything about dogs, and he would have not had, however, time to spare with such duty. Evie is, therefore, entrusted to Johnny's mom, Millie. But, since Millie is busy taking care of the other puppy of the house, the newborn Dickie, son of Johnny and his wife Megan, Tom, Millie's companion, is charged to actually deal with the dog. However, in order to help Johnny in such a difficult time, Frank gets committed to supporting baby Dickie economically, by taking care of him as an "uncle" and visiting him from time to time.

During one of the first visits at Johnny's mom, Frank meets the dog Evie for the first time. The two like each other immediately. Their relationship, mediated by Johnny's family, becomes progressively more intense. Frank asks to meet Evie more and more, to take it out for walks and even to keep it with him each time for longer periods. Millie and Tom, Johnny's parents, get annoyed of such behavior, suspecting that Frank wants to take the dog away from them to earn over its reproduction. The tension between Frank, on the one hand, and Tom and Millie, on the other, progressively increases to the point of leading them to interrupt their relationship, precisely because of his pretensions on the dog. Despite the open hostility of Johnny's parents, Frank will use all his power (economic, of status, etc.) to get the dog free from their power. He will succeed in his goal, at the end of his lover's detention, when Johnny, as legitimate owner, finally on the loose, resumes possession of the dog. But Johnny's return to family life and to his marital obligations has the effect of separating him irretrievably from Frank and Evie, since his inability to devote the necessary attention to his lover and to the she-dog will drive Johnny to sell Evie to Frank, letting the two leave forever.

As can be seen from this short summary, the architecture of the tale is linked to a basic narrative program: freeing the dog from the power of Johnny's parents. It's clear that, once accomplished, this enterprise will radically transform Frank, who will ultimately be relieved of his emotional dependence on Johnny and his bulky family. Making *tabula rasa* of any legacy of his liaison with Johnny will allow Frank to get involved completely in the relationship with his pet, disengaging it from any reference to the past. In hindsight, Evie's liberation program will prove to be instrumental in serving a further existential realization of the Frank + Evie couple, which will be investigated in detail in *My Dog Tulip*, the other novel dedicated by Joe Ackerley to his dog.

3 A Matter of Title

There is a first issue that emerges from the title of the novel, *We Think the World of You*. It can be seen how such an idiomatic sentence is, in the novel, used exclusively by the group of proletarians of Johnny's family who, lacking a proper vocabulary, do not despise to appeal to phrases such as this to express their feelings, to the detriment of any nuance and context. More than once, the narrator distances himself from this attitude that seems to him hateful and essentially hypocritical. To the extent that this expression becomes, however, the title of the novel, an evolution may be reported. From being a stratagem to mark the distance between the narrator and the small world of the characters of the London industrial proletariat, it extends its semantic spectrum up to include the point of view of the narrator as well. In the story, this idiomatic expression is, in fact, not by accident, used in the singular, generally, addressed by one of the proletarians to Frank or to the dog, indicating intensity (e.g., "Johnny thinks the world of you" meaning /cares for you a lot/).

In the title of the novel, there is a "we" that reconfigures the sides: no longer proletarians against bourgeoisies, but rather humans (proletarians and the bourgeois look of Frank) toward animals. In this transformation, the title in question, *We Think the World of You*, ends up not only indicating intensity but also variety /we (all of us) think the world of you (dog)/, each in his own peculiar way. The tale will mark a progressive sliding from the first perspective (proletarians vs. bourgeois) to the second (humans vs. animals) one. Toward the dog Evie, hence, a shift may be noted, which leads the reader from a univocal and generic attestation of affection to the acknowledgment of the various forms which *living together* between humanity and animality may assume.

4 Puppies as Isotope Connectors

As previously said, Frank, Ackerley's alter ego, receives the proposal of his lover to take care of the puppy during the period of detention:

> [...] "*Evie*?" I was astonished. "Who on earth's Evie?"
>
> "You know" said he reproachfully. "She's my dog. Don't you remember? I showed 'er to you last time you come."
>
> I vaguely recollected then that, a month or more ago when, unwilling to assist him, unable to give him up, I had paid him one of those all-too-frequent visits to find out what happened to him, what had happened to me, I had inadvertently trodden, in the darkness of his passage, on something that had squeaked and taken refuge elsewhere. If he had shown me the animal afterwards I had no memory of it. I had been in no mood to notice trivialities. (p.6)

In his faded memory, the dog is "something," an object world like another, a "triviality" with which it is not worth spending too much time and attention. He, as we have seen, refuses to take care of the dog and chooses, instead, to contribute to maintain the newborn Dickie, supporting him economically and going to visit him, from time to time, at Grandma Millie's home.

In the excerpt just reported, an opposition between space models begins to take form. Frank refers to Millie's home – where Johnny also lives – pointing out the discomfort of moving in the "darkness of his passage." The darkness and closure of the spaces inhabited by the proletarians give Frank a sense of oppression and captivity, recognized by difference with the welfare procured by his life led in his bourgeois house, made up of spacious and airy spaces. Proletarian spaces are forced and tight, often closed and claustrophobic, Frank's house and the environments chosen by him to spend free time have in common the peculiarity of being wide and open, settings for the emergence of the self without external constraints.

This basic opposition between spatial and proprioceptive statuses will, lotmanianly, be internally articulated, forming areas where the feeling of well-being or oppression is more or less intense depending on the degree of aperture or closure of the rooms according to a semi-symbolism, so that open: closed = well-being: oppression. These spaces, in which the whole story will be set, may be considered the measure of the identity and difference of the characters involved. See, as a summary, Table 1.

Table 1 Opposition between spaces in the novel

BOURGEOIS SPACES		PROLETARIAN SPACES	
OPEN		CLOSED	
WELFARE		OPPRESSION	
–	+	–	+
Frank's house	**Walks in the parks**	**Millie's house**	**Jail**

Duty of the puppies in the story will be "starting up" the intricate plot of relationships just described, connecting these different universes

5 Dickie vs. Evie

For Grandma Millie, nephew Dickie is the little master of the house. He is reverenced by her as a son, almost, *in place of* her son in jail. On the other hand, Frank, for the whole family, is immediately Dickie's uncle. He struggles to recognize the traits of the beauty that once made him fall in love with Johnny in the rotten features of the newborn. In vain: at his sight, the man cub shows all its monstrousness, crying his eyes out. It is precisely during a visit to his little nephew that Evie emerges unexpectedly from the backyard of the house. And it's a proper discovery:

> She was certainly a pretty bitch, a few months old, rather large and long-legged, and lavishly affectionate in the fawning, insinuating way puppies have. [...]
>
> Evie had now clambered on my lap and was smothering my face with licks.
>
> "Dickie should take a lesson in greetings from this," I said laughing and trying to protect my mouth.
>
> "P'raps she thinks you're Johnny."
>
> Dear Millie, She often made remarks like this which thrilled me to such an extent that they had upon me almost a physical effect. To be identified with Johnny! (p. 23)

Interesting is the double correspondence, which, in these few lines of debut, is indicated: on the one hand, Evie is compared, negatively, to the other puppy (human) of the house, Dickie, while, on the other hand, by means of the same she-dog (which, according to Millie, would not be able to distinguish them), the affinity between Frank and Johnny is stressed. The two puppies are, hence, specularly characterized: Dickie is crude and chubby, completely different from his antagonist Evie which emerges as a perfect puppy, *carino* (Gomorasca 2001) enough in its light beauty. But, there is more. As seen, Evie cannot distinguish Frank and Johnny, since they are substitutable to its canine eyes. Evie ends up assuming, in its fairy-like beauty, the resemblance of interiorities of Frank and Johnny, highlighting the authenticity of their relationship. The little she-dog becomes, more and more clearly, a complex term of the union between the two lovers, embodying their identity. But, to see, things may even be more complicated. Evie is a *vague* sign of continuity between the identities, actually defined, of the two lovers, now emerging as Frank's alter ego, now as the jailed Johnny's avatar or even as a synthesis of their union. In this case, the dog turns into a hybrid and instable subject, fragile and enigmatic symbol, which is hardly completely revealed, hardly reduced to a univocal meaning.

Dickie, on the contrary, takes on the traits of the evil puppy (see Surace 2017). Frank considers Dickie's biological offspring from Johnny as the outcome of his surrender to the expectations of the heterosexual bourgeois world and, on the other hand, as a sign of his class affiliation. Dickie becomes a figure of all that is irreconcilable between the bourgeois and the proletarian world. Two parallel processes of kinship-building may be here recognized. On the one hand, there is the institutional way, represented by the *man cub* Dickie. This first way, as said, is seen by Frank as the humble abdication of his gay companion to the demands of his social group (marriage, children). The outcome of such a conduct can only take the form of disgust, at his eyes. On the other hand, there is Evie, the only way out of standardization, just because of its paradox. Its status of pet, commonly considered as a lower form of life, and therefore incapable of acting as an actual "relative" in the family, will stand as a sign of the troubled love of the two men.

To the biological continuity personified by Dickie, the two impossible lovers Frank and Johnny will prefer the bestial *elective affinities* with their canine double. It is no coincidence that only the two lovers acknowledge the dignity of being a subject to the puppy, whereas the rest of the family members consider Evie as a mere object, a bearer of economic value in view of its upcoming ability to reproduce. The fact that the dog's subjectivity is not acknowledged by the most, that its existence is, in the end, considered irrelevant, dispenses it from the obligations of affiliating in the shady world in which it is also forced to live and will constitute for the dog a warranty of alterity. Its marginalization will, in short, represent its salvation.

The little puppy Evie acts, therefore, as a *totem*, at the same time, proper body, individual, and flag of the specificity of the existential bond that links Frank and Johnny.

6 Tamings

Tom, Millie's companion, and Frank are carriers of two different visions of the dog. For Tom, Evie is a mere economic mean, a reproduction machine upon which to gain. On the contrary, the same dog is considered as a full subject by Frank, a bearer of meanings related with identity, a totemic double of the existential state that links him to his lover Johnny. From this underlying opposition of values over the pet (with Floch 1990, 1995 – *Utopian* vs. *Critical*), a series of consequences does follow. First of all, a divergence on what kind of education providing it with emerges. To the extent that Johnny's parents agree not to recognize a complete subjectivity to the dog, they choose to reserve the caring of the only woman of the house, Millie, to the legitimate heir, Dickie, leaving the management of the dog to Tom. The way he will proceed in this task is significant: his approach to education reveals strongly authoritarian, linked to a coarse stimulus-response setting (how to forget that at the base of cognitive theories there is a dog[4]?), made of physical violence and constraint.

Already since the onset of the dog on the scene, in the passage mentioned above, the two antagonists Frank and Tom have been applying different forms of interaction (Landowski 2005) toward the quadruped:

> "I've just worming 'er," said Tom, who had resumed his habitual arm-chair by the fire.
> "Oh," I said politely. "Did she have worms?"
> "All puppies 'as worms. They're born with 'em."
> Evie had now clambered on my lap and was smothering my face with licks.
> [...]
> "Do you know what's good for worms?" asked Tom.
> "No," I sad repressively.
> "'Uman' air chopped up in treacle. My old grand-dad told me that. That wasn't much 'e didn't know about dogs..."
> "Now leave me alone, do!" Millie exclaimed to Evie. Or I'll have to put you out!"
> " 'Ere, Evie!" called Tom, and she scrambled up on to his knees and began to lick his grey, cadaverous face.
> "Tobaccer's good, too. I 'ad a dog once what ate up all the fag-ends in the street. If they was burning 'e'd stub 'em out first with 'is paw. 'E kept 'isself free from worms...." (p. 23)

While Frank feels overwhelmed by the puppy's affectionate vitality, Tom's commentary only concerns the physiological side, in an effort to minimize the immediate sympathy between the two and to show off his expertise on the dog's body machine, not avoiding to get into disgusting details.

It is no coincidence that, in the meanwhile, Frank looks over an old picture of Johnny as a young cadet:

> It had been taken during his insubordinate career as a sailor, and I gazed up at it again with a pang. How attractive he had been with his short, strong, lightly balanced figure and springing gait. The whole shaft of his beautiful neck, his wide shoulders and deep chest, his narrow hips, everything that he had had been almost effeminately displayed by that extra ordinary close-fitting costume of ribbons, bows, and silks. And what fun he had been, so lively and so gay... Much water, alas, had flowed under the bridge since then. A momentary weak feeling of anger and self-pity took me at the thought of what he had been and what marriage had done to him. Boy though he still was to my older mind, the straight back was now a trifle bowed, but the face was the same and from behind the glass of the picture the limpid eyes looked down at me steadfastly, as reassuringly as they had looked from behind the glass of the visitor's booth. He had promised a happier future; they seemed to confirm the promise. It was with a feeling of refreshment that I returned to the kitchen. (pp. 24–25)

Johnny's beauty is made of orthogonality, vertical lines, and sudden rushes, but it is also tactile and lies in the delicacy of the tissues he wears. His distinct elegance characterizes him as an aberrant creature with respect to the prosaicness that the narrator recognizes to the other members of Johnny's family and social class. This same elegance represents, in Frank's eye, his nobility, which perfectly combines with a libertine and light-minded irresponsibility. All of this is destined to be

[4]Reference here goes to the known experiment of the Pavlovian conditioning (Pavlov and Anrep 1927).

mortified by the marriage, a path that is, in all respects, prefigured as a process of taming, here meant as the denegation of any primitive, noble, and pure spirit of insubordination. The signs of such a treatment will be well visible on Johnny's body, destined to be deformed by the assimilation in the proletarian world. The prison, which, at a certain point, Johnny will have to deal with, only represents the logical continuation of this process of progressive reduction of autonomy.

Without Frank's intervention, the dog would have met the same fate. The Alsatian shepherd's puppy thus becomes a figure of the unhappy fate of Johnny, doubly, either because it has to give up his nobility by adapting to the rude and sloppy proletarian life or because it is "imprisoned" as his master, in a house where no one takes care of it properly. The taming of the dog includes the denial of every of its animal prerogatives, of every resistance of subjectivity. Forced as it is to live in a cramped and unhealthy backyard and beaten at every opportunity, even the evening walks, so vital for a dog of its size, are denied to it. Such a lifestyle would have soon caused irreparable damage, turning the dog into dysfunctional and aggressive. This condition is prefigured by the feeling of claustrophobia experienced by Frank during his visits to Millie's home. It will be precisely this feeling to mark the irreducibility between bourgeois and proletarians:

> I noticed at once that, in my short absence, the window had been closed. The working classes, I reflected with a shrug, have an ineradicable belief that the colds from which they constantly suffer are due to fresh air rather than to the lack of it. With such superficial mental comment did I dismiss an incident to which, I realized, when I recalled it some months later, I would have been wiser to pay more serious attention, for it set in a way, the psychological pattern of much that followed. (p.25)

The feeling of suffocation, of constraint, and of forced resignation of the most elementary (like breathing) demands represents, as already stated in the opening, the underlying isotopy of the tale and marks the deep identity of the characters, each one, as we have seen, in his own way, forced, in the humble proletarian life. But the constriction of the puppy becomes not only a figure of the two lovers as individuals but also of their love story, of their couple's condition, obliged to make a good face to bad play, to adapt to rules of a world that simply does not imply that their feelings might find a socially acceptable form (Table 2).

Obviously, such a *status quo* requires a reaction that is already taking form since the first walk. Straightway, empathizing with the dog's feeling of choking, after having seen it imploring Tom to be taken out (thing which, like all dogs, Evie does pointing at the leash), Frank blurts with a liberating "I'll take her" (p.35).

Table 2 Evie as a complex term

	Evie	
Frank	**Johnny**	**Frank + Johnny**
Closed windows	Marriage Imprisonment in jail	Inability to be accepted

Unlike Dickie, the other puppy at home, Evie accepts Frank's friendship with enthusiasm, walking with him toward the first steps of their tiny little revolution. The emotional climate is now totally different from the walks that were rarely allowed by the proletarian family. The dog wants to run. Frank, rather than following Tom's advice of pulling the collar, decides to loosen it, following Evie's demand for freedom. This gesture has, as a first consequence, the emergence of Evie's will. Before running freely on the alley, Evie thinks a little and, in any case, by deliberate choice, doesn't go too far away from its new companion, perhaps – Frank infers – for the fear of getting lost. During this short period of time, there is also the opportunity to do another forbidden experience: playing with another puppy, simply because "dogs enjoy to talk to one another" (p.39) where what the dog might enjoy was not of interest for any of its caretakers before.

Once again, the emotion for the beauty and freedom experienced by the two new walking comrades emerges by difference with the disgust provoked by Dickie. When back home, Frank finds Millie dealing with the little Dickie, discovering that he had picked some butter from a plate and smeared it in his face. If, during the walk, the emergence of the instinctive side of the bitch creates emotion and transport, Dickie's mischiefs, accepted with love by his grandmother, reveal ridiculous and disgusting at Frank's eyes.

7 Pedagogical Models in Comparison

Like every story that deals over the relationship between humans and pets, also *We Think the World of You* comes as a moral tale, setting up a clash between forms of interaction, between pedagogies. It is precisely in the story of the walk that, apart from the difference in educational models, a clash between addressers emerges. On the one hand, there is Tom's position, assimilating Evie's condition to that of an object to be programmed. On the other, stands Frank's one. He tends to make no prediction about the dog behavior, taking the risk of leaving it free, confident that the freedom granted would have let the "wisdom"/subjectivity of the dog emerge from the prostration. If Tom's goal is to break any dog's autonomy by reducing it to submission, Frank's "no-rules" education is functional to the reestablishment of Evie's original canine pride, meant as emancipation from its state of obedience. The semiotic square can effectively describe these dynamics of power, so central in every story about puppies. With Greimas (1983), see Table 3.

The conflict with Frank emerges at this point. He wants to drive the dog to recover its lost pride, by guiding it back to the sovereign state of freedom, which imply a free and transparent *can do*.

Frank and Tom, carriers of the opposite pet conceptions, reveal so distant that Tom cannot even imagine Frank's model being taken seriously. According to Tom, Frank's showing of his passionate love toward Evie is, in fact, a cover for his economic aims over it, intended at earning on its reproduction. Not understanding the underlying rationality of Frank's behavior, Tom tends to think of him as an *incognito*

Table 3 Semiotic square of power articulation

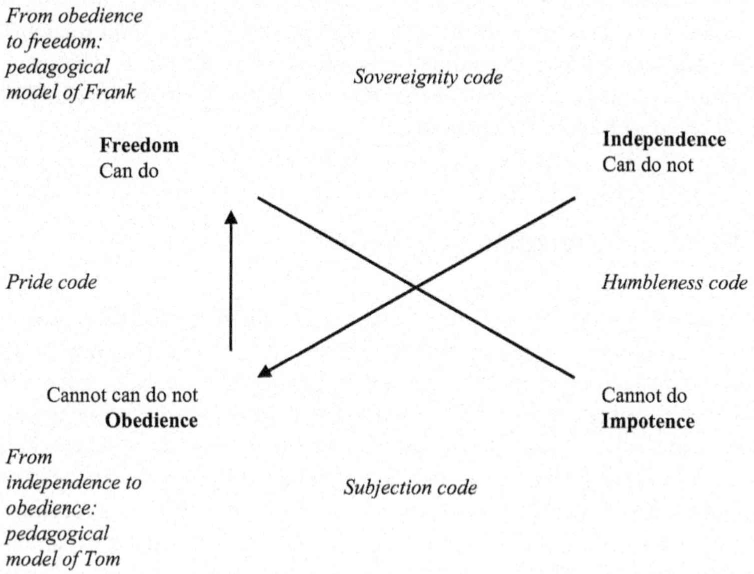

agent, a malignant antagonist secretly playing at his own level of value, that is, the exploitation of its fertility for economic purposes.

Right from this conjecture, Johnny's family will begin to marginalize Frank, considered guilty of willing to appropriate of the puppy for his shady business while neglecting his role of Dickie's caretaker. So, again, the problem of respecting the social order comes: to keep in contact with his beloved she-dog, Frank is forced to barter it with Dickie. Supporting him economically and showing some kind of concern with him, despite his ugliness, are, for Frank, the only ways to see his dog of choice.

In the constant comparison that Frank leads between the human puppy, Dickie, and the canine one, Evie, a disparity, a real injustice, emerges. Dickie is the household little prince; he stands in the existential condition that, in the Greimas square quoted above, may be referred to as the "code of sovereignty," realized by the double possibility of *can do* (as when, unpunished, he fills his face of butter) and that of *cannot do* (not honoring the gifts he receives from Frank, while bursting, instead, duly, into tears). However, Dickie holds this position undeservedly, by pure descent, emerging as a subject without any quality and talent. A situation similar to the one described in another memorable tale focused on the relationship between humans and animals, which is the Pixar movie, Ratatouille (see Mangiapane 2014). Even in that story, there is a human, privileged and inept (Linguini) affiliated to the *in-group* just because of biological continuity (*ius sanguinis*). On the other hand, a *mouse chef* (Rémi) asks for citizenship in the name of its skills and its ability to contribute to the life of the community (*ius soli*). Although the moral lesson of the film, "everyone can cook," stands on the side of the mouse, its acceptance must be kept hidden to the social body, in order to be actually realized: the mouse can certainly cook but has to do it in secret. Such a *status quo* cannot be accepted by Frank who is in an existential state comparable to that of his canine double: his homosexuality is, in fact, regarded as a detrimental trait in Puritan England in the 1950s. The struggle for the liberation of his pet appears, then, as a mirror on the symbolic level of his battle to be accepted as Johnny's companion.

8 Walks and Emergence of the Self

As soon as the dog and its aspiring master go around, their friendship grows, made of long walks and adventures around the suburbs of London. This tireless wandering, indulging the dog's most bestial needs, creates a scenario of deep empathy, giving shape, in fact, to a sort of extended body, *dual*, made of an increasingly intense consonance, obtained through a mutual adjustment (Landowski 2005). While it is true that thanks to the freedom guaranteed by its new tutor, the dog can bring out its basic docility in spite of any animosity acquired as a reaction to the hardness of its life at Millie's house, a specular feeling will be experienced by Frank: not only the dog learns from the master, but, above all, the master learns from the dog to reveal his emotion, his most secret and disguised character, awakening it. At play in this game is the sense itself of life, which can be retrieved in the relationship with the pet, as emergence, as a reaffirmed truth:

> Now I had the pleasure I had promised us both, the pleasure of setting her free upon the grass. And her reward was mine. Across the open spaces of the park the rough wind blew with its full strength, and she became a part of the dancing day, leaping and flying among the torn trees, wild in her delight. And her gratitude was as boundless as her happiness. The same watchfulness, the same invitation, that I had already noticed in her governed her behaviour still. Gay though she was, it was a shared gaiety always; to caper about was not enough, I must caper too; and who could have resisted such ebullience of spirit, which caught one up into itself and buffeting wind? (pp. 88–89)

This excerpt describes a peculiar dynamics that can be understood as a proper *saisie estétique* (Greimas 1987). It is a moment of ecstatic suspension, effect of an extra-narrative truth and, therefore, impossible to be told but by approximation, only once over, in respect of which assuming an attitude of loss, of nostalgia. It is at this point that the construction of a syncretic, at all respects, hybrid (Latour 1992), actant occurs. We might define this actant, obtained through the degradation of the subjectivities of both dog and master toward a common sensibility, *dual*. See the semiotic square in Table 4.

The realization of this dual subjectivity, the result of the *saisie estétique*, is, as stressed, elusive, similar to an epiphany. And it is the nostalgic attitude toward it that drives dogs and masters to mold themselves so that it can take place again. In their daily lives, they do nothing else but model each other, try to make attempts of similarity, often even paradoxical, with the ill-conceived hope of reviving the revealing leap of the *saisie estétique*, which would lead them to experience their *dual* identity

Table 4 Construction of a *dual* actant

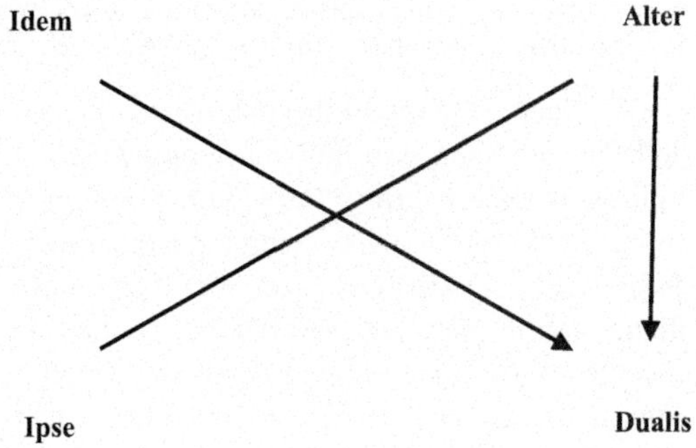

Table 5 Levels of fusion between subjects and world

Third level (second connection)	Frank + Evie + natural world	
Second level (first connection)	Frank	Evie + natural world
First level (disjoined actants)	Frank	Evie

once again. But as we know (again with Greimas 1987), the *saisie estétique* can only reveal as *imperfection*, a radical manifestation of the unexpected.

Getting back to the brief excerpt above in the view of these remarks, you can see how the isotopy of liberation staged throughout the story requires a very precise setup which is a necessary condition for the leap of the *saisie estétique* to take place: the open space, the chance to run free on the lawn make the dog one with the wind blowing on the trees, even them animated too, of the park. The constitution of such a complex environment forms a totalizing image capable, in the first instance, of revealing the greatness of Evie's femininity as part of a high and sublime natural world and, therefore, of letting Frank, invited by the dog to join, access this awareness. It is at this point that Frank and Evie's identity succeeds in reaching an even higher level, melting in an ulterior natural whole (including trees, wind, etc.) (Table 5).

Assuming this perspective, any walk held by dog and master may be considered as a way of preparing the scene to the miracle of such superior perception. Since the conditions of the "triggering" of the *saisie estétique* are unpredictable, the aspiring subjects, nostalgic of it, adopt a kind of philological position, aiming at reconstructing the scenario and the dynamics of the first trigger, in order to create the conditions for it to happen again. Not all the walks had by the two protagonists, therefore, turn into unforgettable moments of realization of duality, but, for this same realization to occur, two subjects need to spend a lot of time together, in open spaces, perhaps walking.

But there is no real *saisie estétique*, without change:

> The daylight hours we spent mostly in the open air. Evie saw to that. And it was borne in upon, without perceiving it, I had grown old and dull, I had forgotten that life itself was an adventure. She corrected this. She held the keys to what I had lost, the secret of delight. It was a word often used, but what did I now of the quality itself, I thought, as I watched her

inextinguishable high spirits, her insatiable appetite, not for food, in which she seemed scarcely interested, but for fun the way she welcomed life like a lover? (pp. 93–94)

The deep meaning of the *saisie estétique* is, for Frank, the awakening of his own self. The attendance of the dog-puppy wakes in him, aged and senile bourgeois, the willing of rediscovering "the secret of delight" which is having fun, welcoming "life like a lover," and rediscovering the exuberance of his lost youth. We are once again faced with a peculiar characteristic of the puppies, that is, putting the human/animal distinction in crisis. The exuberance of the little ones and their ability to awaken in their tutors (whatever they are parents, grandparents, or masters as in this case) the impression of having awakened, thanks to the time spent together, their most genuine spirit, purified by any necessary compromise of adult life, thus rediscovering their deeper and better identity. Elsewhere (Mangiapane 2015), I recognized such dynamics with roles reverted: in the lucky series of cartoons targeted to the early childhood, Masha and the Bear, the young and very human Masha succeeds to revive in the lazy Bear the lost playfulness of its youth. The arrival of the puppies in a group that is settled undermines its dynamics and forces the tutors in charge to "adjust" (we always refer to Landowski 2005) responding to their requests. But this turmoil, often prefigured and hence initially opposed by the parties involved (it makes no exception our *We Think the World of You*, where Frank, at Johnny's request to deal with Evie during the detention period, opposes a firm rejection), works, in all respects, as an invitation to seek lost time, to liberate from all the frills and the heavyweights of cultural mediations in the name of the rediscovery of the childhood self, forgotten and buried poetic double but yet custodian of the authenticity of the self. The regression, result of this adjustment, ends up to determine peculiar forms of interaction: fluttering, twitching, and infantile play between tutors and puppies, dogs and masters, are usually so bizarre and impenetrable to look foolish at the outsider's eye but, after all, can be effectively studied (as Battistini 2017 does) as a characteristic sign of the relationship between adults and puppies.

9 Second Saisie Estétique: From Animism to Analogism

At this point, Frank is tempted to look for a deeper similarity with the dog, based on a common point of view on the world, despite the differences of species. It happens immediately after:

> There was, however, one quiet little indoor game with which, on the second evening, she entertained me and, to speak the truth, momentarily disturbed my mind. Not that I could have retreated then, even had I wished to retreat, I say to myself looking back; I was already too deeply involved. In any case the whole thing could easily be laughed off… It began simply enough. She was sitting on the divan facing me, staring at me, her long forelegs close together, the paw joints flexed over the edge of the bed. Sitting thus, she suddenly picked up her ball which, with various other objects close together, the paw joints flexed over the edge of the bed. Sitting thus, she suddenly picked up her ball which, with various other objects to which she seemed to attach a value, she had collected about her, and set it

on her legs. It rolled down them, as upon rails, fell to the floor and bounced across the room towards. This was nothing. Pure accident. Merely amusing. The mechanics were easy; our relative positions directed the ball inevitably from her to me. Receiving it into my hand I returned it with a laugh. She caught it in her jaws. But then she set it on her legs again; down them it rolled, bounced across the carpet and reached my hand. Now I looked at her with more particularity and put my book away. The ball was in my hand and she was gazing at me expectantly. For a second I hesitated, as though a cautionary hand had been laid upon me. Then I cast it back into her waiting jaws. She placed it upon her legs a third time. It did not move. Peering down at it, as if in perplexity, she gave t, with her long black nose, a shove, and it began once more its slow conversational journey from her to me. But no, just as it reached the verge –was t simply because she childishly felt she could not after all bear to part with it, or because the hitch that had occurred had vexed her? –she suddenly seized it back with a swift, almost scolding, thrust of the head and replaced it on her legs. It rolled. It fell. It bounced. It crossed the room and came into my hand. Yes, yes, of course, I know; it is absurd to read too much into animal behaviour, and afterwards, as I have said, I laughed it off; but at that moment I did take the uncanny impression that, in deliberate and purposeful way, she had gathered up all her poor resources and, in order to reach me directly and upon my own ground, had managed to cross that uncrossable barrier that separates man and beast. The expression on her face contributed to this fleeting illusion. Some animals have a furrow above their eyes very like that furrow, etched by a lifetime of meditation that we see upon the brow of sages. In the animal's case, of course, it is merely the loose skin wrinkling upon the line of the socket bone; but it often imparts to their faces a similar look of wisdom. Evie has this "intellectual" line, and it lent to her expression now an appearance of the profoundest concentration. With her nose pointing down and her ears cocked forward she followed, with the utmost gravity, the progress of the ball as it traveled down her legs, fell over the edge, bounded across the carpet and reached my hand; then, without altering the bent position of her head, she raised her eyes beneath their furrowed brows to min and directed at me the kind of look that two scientists might exchange after successfully bringing off some critical experiment in physics. Yet, when I returned the ball to her now, it was as though the effort she had been making–if effort it was–suddenly failed; she became a mere dog once more, kicking up her legs and rolling about with the toy in her mouth; and when I offered out of curiosity, to replay the game next evening, I could not get it going; she seemed worried and confused; the inspiration, having done its work, had apparently gone out of her forever. (pp. 96–98)

This excerpt results very interesting for various reasons. Firstly, it acknowledges a fundamental feature of the stories focused over the relationship between men and animals, which is the fact that they require a precise enunciation style. It is a phenomenological, approximated, sensitive glance that is intended to describe the minimal significant scraps at hand useful to recognize subjectivity to the animal. This attitude corresponds to a clear choice: those readers who want to understand the meaning of the story and the particular relationship between Frank and his Evie have to pay attention to details that, seen from outside, might seem minimal and insignificant. This is why the piece stages two points of view, one inside the relationship with the pet and the other outside. The whole episode, as well as the whole novel, is within an enunciative *debrayage* that clearly separates the self-narrated by the narrator. Here, in particular, disagreement emerges. The self-narrator, located in a very faraway time from the facts being told, assumes the position of external evaluator of the story:

> Not that I could have retreated then, even had I wished to retreat, I say to myself looking back; I was already too deeply involved. In any case the whole thing could easily be laughed off…

The difference between the two subjective instances, what has changed in the meantime, is the involvement. Being inside or outside the relationship has consequences on the ability to judge this same episode. Seen in retrospect, what is told appears ridiculous to the narrator. On the other hand, the fact itself that this same episode has been chosen and told despite the time elapsed says a lot about its pregnancy.

The difference of judgment between these two subjective instances is at the heart of the dramatization. The work of the heroes onstage is, therefore, surely evaluated from outside, after time, but also in the meanwhile, in the interpretation that dog and master give of it while in the situation. These different assessments coexist, so that the narrator criticizes the judgment given by the subject involved:

> The expression on her face contributed to this fleeting illusion. Some animals have a furrow above their eyes very like that furrow, etched by a lifetime of meditation that we see upon the brow of sages. In the animal's case, of course, it is merely the loose skin wrinkling upon the line of the socket bone; but it often imparts to their faces a similar look of wisdom. Evie has this "intellectual" line, and it lent to her expression now an appearance of the profoundest concentration.

Narrator subject's position stands for a conscious appraisal of the irreducible diversity between dogs and masters (*analogism*), motivating and explaining the terms of the aberrant interpretation of the *embrayé* subject, whose considerations tend to represent, on the contrary, the whole scene in a perspective that we might call, with Descola (2005) *animist*, recognizing, despite the differences of physicalities, an *almost-humanity* in the subjectivity of the dog. The gap just indicated implies a repositioning of figurative traits. The *embrayé* subject explores the dog's body by difference. Evie's body is presented as a *rebus* to be deciphered, by looking for some resemblance able to legitimize the aspiration to identity. Thus, the merely visual features of the dog's wrinkled skin under the orbits end up being guessed as an "intellectual line" capable of legitimizing the identification of the dog as a *quasi-human* thinker. The narrator will stress the mistake: these same characteristics, in the later reading of the facts that is offered, will be degraded to an iconic characteristic of Evie's animality. But, such an assertive iconic recognition of the "bestial" trait, made a posteriori, cannot completely persuade neither the narrator nor the reader. Though explicitly speaking the narrator's position is not discussed, his hesitation aims at producing an appropriate effect of "it's not true but I believe it" on the reader, which emerges from the very choice of telling that episode rather than another, from the space that is granted to the episode itself, from the almost investigative approach, with minute and accurate descriptions, which characterize the excerpt. To be dramatized, among overlying enunciation regimes is thus the undecidability over the attitude to keep toward the dog, the uncertainty of the ideological position to be taken. This anxiety is a specificity of the stories about pets. The experience of life with pets ends thus to form a pool of *animistic* objections to the dominant paradigm. Objections of this kind are mostly lived as *counter cases* which are not able to scratch the ideological horizon of reference; even if they are confusing, they are able to undermine the epistemic certainty of the "involved" subjects. These kinds of characters, master's *embrayé* in the relationship with their pets, thus become witness of a proper ontological challenge. Uncertain on how to behave with the *alterity* of their nonhuman life companion, they emerge as translators/traitors of ontologies to their neighbor.

One can observe how these scraps of ontologies appear as poetic experiences. Even in this passage, a *saisie estétique* emerges that reflects the identity of the subjects at stake, dog and master, in a radical way. The dog will be able to reveal, even if it is only for a moment, friend, *colleague*, equal. In this new condition, the puppy itself takes the floor as a judge of the incident:

> With her nose pointing down and her ears cocked forward she followed, with the utmost gravity, the progress of the ball as it travelled down her legs, fell over the edge, bounded across the carpet and reached my hand; then, without altering the bent position of her head, she raised her eyes beneath their furrowed brows to min and directed at me the kind of look that two scientists might exchange after successfully bringing off some critical experiment in physics. Yet, when I returned the ball to her now, it was as though the effort she had been making–if effort it was–suddenly failed; she became a mere dog once more, kicking up her legs and rolling about with the toy in her mouth; and when I offered out of curiosity, to replay the game next evening, I could not get it going; she seemed worried and confused; the inspiration, having done its work, had apparently gone out of her forever. (p. 98)

Again, ephemeral and elusive, the *saisie estétique* leaves the subjects involved with a sense of restlessness, but what really matters is that the time of mutual recognition cannot last any longer, releasing the two protagonists of such extraordinary experience to their differences, which, at this point, beyond doubt, also appear on the side of the interiorities: not just bodies but profoundly different souls.

This awareness, gained through the frequent mutual attendance, is at the core of the relationship between the master and his puppy, leading them to the acknowledgment, at the same time human and canine, of the unattainability of a similarity of interiorities despite the species differences. The subject "involved" Frank seeks out, with Descola (2005), the identity of interiorities despite the differences of physicalities, embracing a conception of his relationship with the pet as an *animistic* digression of a life based on a fundamentally *naturalistic* paradigm. The failure of the delicate and poetic experiment just reported will guide the parties to a new awareness: not fellowship of souls despite the diversity of physical form but symbolic efficacy, sign, and semiotics.

The love story between the dog and his master will henceforth look for a replenishment of Evie's lost animality: the more Evie will get back to be animal, to regain its own body, its physical needs, its sexual appetites, its own temporality, and its

Table 6 Ontological shifts

Dominant paradigm	Naturalism			
Ontological shifts in the relationship with puppies	Naturalism	Totemism	Animism	Analogism
Evie	Refusal of accepting *Evie*	Solidarity as outsider	First *saisie estétique*	Second *saisie estétique*
Dickie	Dickie holder of subjectivity	Traditional family		

peculiar way of managing the affections, the more, in short, the difference between Frank's human form of life and that of his puppy will be pointed out. The lives of Frank and Evie become, in their dissimilarity and verified irreducibility, similar and, hence, translatable, expressions of a profound solidarity that gets formed, over time, precisely because of the difference. This is, always referring to Descola (2005), the position of *analogism*.

It is worthwhile, at this point, returning to the comparison with the detestable grandchild with whom, constantly, the affection toward the little puppy Evie has been compared. While Frank, as a compensation for his role as Johnny's lover and supporter of his economic condition, is offered to act as an acquired *uncle* for Dickie, he accepts this role with reluctance as the newborn represents a model of affiliation and affection whose shape cannot be accepted by him, except at the cost of humiliation. For this reason, no complicity may be triggered with the other puppy Dickie, no emotional bond with him, since he becomes, in Frank's vision, figure of all that is hostile. Obviously, because of this incompatibility, their relationship will not evolve as it does with the other puppy, Evie, intended to live with Frank for a long time.

We can thus recognize very precise sequences in the evolution of the story and the relationship between Frank and the two puppies of the story (Table 6).

If the beginning of the story marks the refusal of Johnny's request to deal with Evie in the name of a reasonable consideration of the constraints that dealing with a dog would have implied, the recognition of a subjectivity to the puppy goes through a first identification: the dog reveals itself as a perfect totem, an existential model of the specialty of the relationship between Frank and Johnny and their form of life. As long as Frank and Evie get familiar, things will change. More and more, the two protagonists will "adjust" each other, assimilating rhythms and routines, tastes and desires, up to experience a full identity despite their difference of physicalities. During their walks, they will melt in a kind of emotional fusion, assimilating themselves completely while, at the same time, aspiring to greater communion. Thanks to Evie's encouragement, Frank, even though only for a moment, will be able to *become animal* (Deleuze and Guattari 1980) but even more radically *become natural*, experiencing a further belonging to a natural allure involving as a one-off, stones, wind, trees, animals, and humans. But such an ambition of "involved subjects" will collide with the subsequent acknowledgment of an essential irreducibility of the two worlds, human and animal. This will lead to a shift of strategy. Rather than thinking themselves as twin souls in disguise, from that moment onward, every effort of Frank will be focused on individualizing his relationship with Evie, so that the fundamental traits of his character may emerge by difference with that of the dog. This difference will bring the subjects to a further awareness of their identity, to a self-recognition (with Ricœur 1990, *ipse*), through analogy, namely, the finding of profound common existential patterns and similar paths of life despite the difference.

The struggle for the liberation of the dog from the oppressing proletarians changes polarity. From being figure of the relief pursued by the two lovers, it turns into a struggle for individuation, a struggle for the claim of the canine difference and the rediscovery of the true Evie. This is the condition through which the identity and friendship of dog and master, this time meant as singles, individuals, or companions, emerge.

The novel ends with Johnny's release from jail. As expected, Johnny reconstructs his family, reuniting with his wife and children. He also chooses to take possession of the dog, dealing with it in his daily life, getting committed to sharing it with Frank sometimes. The dog, thus, becomes the only way to keep alive the relationship between the two lovers. However, Johnny's spare time will be reclaimed by his wife. She will reveal jealous of both Evie and Frank. Such situation will force Johnny to look for a compromise solution, entrusting Frank with the day-to-day care of the dog. If, at first, the woman agrees to share her husband, by means of the dog, with Frank, soon this precarious balance falls apart. Frank ends up buying Evie from Johnny, who progressively, absorbed by his couple's life, moves away from both.

After a while, the two will meet at the park, each accompanied by his consort. Johnny, with his wife and children, bent and *tamed* with an unhappy marriage, will show a touch of anger toward the condition of his former companion. In his view, among the two, Frank, with his interspecies relationship free of conformism, has been luckier than himself.

10 Conclusions

This contribution tries to investigate the question of how to frame, from the point of view of the affiliation models, the relationship with pets. In this regard, *We Think the World of You* reveals as an exemplary text. The dog/master relationship, therefore, appears constituted of an uncertain status. This same relationship appears to Frank as a loophole to subvert the "taming" reserved by Johnny's proletarian family to all members of the family and as a perfect symbol of the peculiar existential dimension of the subjects at play. So it happens that the struggle Frank triggers with Johnny's family to release the dog becomes a *figure* of his existential condition of outsider and of the forbidden relationship that links him to Johnny. But the emotional relationship prevails on this aspect, opening up to scenarios of multiple ontological discards. The pet will, therefore, be able to awaken, in its aspirant master, a cheerful and primitive dimension buried by the dust of his bourgeois customs. Frank identifies with his own pet, despite the difference of species. Moreover, he will recognize himself, along with his pet, into an expression of a superior natural form that encompasses them both. He and the dog will exactly become "world that dance," as we have seen. But this vision reveals as fleeting and unsustainable, still through to a *saisie estétique*, which renovates again to the relationship within them. The two come to a mature awareness, accepting their difference without trying to reduce it. Their relationship levels up in mutual acceptance of their differences, which will be a fundamental premise for the creation of a more subtle identity, no longer among more or less similar characters but as a promise over time.

The result of this inquiry, with the description of the steps already seen, is an attempt to "deontologize" the multinational model of Descola (2005), showing how it can be useful, as well as to indicate general and tendentially constant ideological paradigms of entire civilizations, also to describe the oscillations between forms of thought, tactically, in the interaction between subjects, in everyday life. The point of view here used tends to hold together both dimensions, trying to highlight the continuities, or conversely, the discontinuities, between forms of community affiliation and localized thinking of individuals. They constitute a competition of forms of rationality, which are selected and made operational within specific strategic scenarios. In short, following de Certeau (1980), it can be said that ontological affiliation represents a form of power "from above" which is normally subjected to a "downward" *re-semanticization* by individuals. Given a dominant "ontological" paradigm – in the case of the analyzed novel, the naturalist position – interactions may put this affiliation in brackets by negotiating spheres of adhesion's suspension. These spheres, however, do not aspire to question the whole paradigm but to surviving as extemporaneous solutions, whose meaning and pregnancy are postponed indefinitely.

Halfway between the institutional ontological form and the individual one, a third, properly semiotic, can be pointed out. It has to do with the possibility of recognizing the creation of proper ontological practices in function of certain discursive configurations. We have seen how puppies, both human and nonhuman, as a thematic role, constitute a discursive configuration that works as a proper access gateway to ontological reconfiguration. The story of Frank, of Johnny, and of their little dog is doubly exemplary in this regard. On the one hand, since it is fiction capable to "socialize" an individual path by transforming it into a "thought experiment" at the disposal of any reader of any time. On the other hand, it is exemplary in so far as it highlights a particular attitude of the relationship with pets of turning into a perturbing element of the established order, into an element of perplexity in front of the other. This oscillation, this incertitude on what status to grant to the other, is the perfect setting for the multinatural clash that arises from the will to resolve the ambiguity, by raising the voice, by cutting the tips on the table, and by reducing the semiotic stake to files already scheduled at post. At one point, among the various attempts to rescue the little Evie from Johnny's parents, Frank asks the owner of a dog's pension for taking care of the dog for a fee, during the months of Johnny's detention. After verifying the irascibility of his Alsatian shepherd since held in captivity too much time, she begins: "Shoot her. It's the kindest thing to do.

I know exactly the sort of dog she is, had dozens of 'm through my hands. She'll never be any good now, take it from me, she'll be a nuisance to you, herself and anyone else, besides giving the breed a bad name, which it's unfortunately got already through mishandled dogs" (pp.90–91). Frank listens to these words without being able to gain from the profound reductionism contained therein. It is pointless to point out that the rest of the story shows that, despite by the hands of the dog-pension owner, dozens of such dogs had passed, she was wrong.

References

Ackerley, J. R. (1956). *My dog Tulip*. London: Secker & Warburg.

Ackerley, J. R. (1960). (2011) *We think the world of you*. New York: New York Review Books.

Battistini, E. (2017). Cariiino! Vezzeggiativi ed espressioni vocali nel parlare ai cuccioli in Mangiapane F (ed) in E/C n. 21 (forthcoming).

Boutaud, J.-J. (2013). Dimagrire in regime mediatico. In D. Mangano & G. Marrone (Eds.), *Dietetica e semiotica. Regimi di senso*. Milan: Mimesis.

De Certeau, M. (1980). *L'invention du quotidien. I Arts de faire*. Paris: UGE.

Deleuze, G., & Guattari, F. (1980). *Mille Plateaux*. Paris: Éditions de Minuit.

Descola, P. (2005). *Par-delà nature et culture*. Paris: Gallimard.

Floch, J.-M. (1990). *Sémiotique, marketing et communication: sous les signes, les stratégies*. Paris: Puf.

Floch, J.-M. (1995). *Identités visuelles*. Paris: Puf.

Gomarasca, A. (Ed.). (2001). *La bambola e il robottone. Culture Pop nel Giappone contemporaneo*. Turin: Einaudi.

Greimas, A. J. (1983). *Du sens II. Essais sémiotiques*. Paris: Seuil.

Greimas, A. J. (1987). *De l'imperfection, P*. Paris: Fanlac.

Landowski, E. (2005). *Les interactions risquées*. Limoges: Pulim.

Latour, B. (1992). Where are the missing masses, sociology of a few mundane artefacts. In W. Bijker & J. Law (Eds.), *Shaping technology-building society. Studies in sociotechnical change* (Vol. x, pp. 225–259). Cambridge, MA: MIT Press.

Mangiapane, F. (2014). Scontri etnici e corpi gloriosi. Mangiare al cinema. In G. Marrone (Ed.), *Buono da pensare. Cultura e Comunicazione del gusto* (pp. 133–166). Milan: Carocci.

Mangiapane, F. (2015). *Il tempo ritrovato di Masha e Orso*. Available via Doppiozero. http://www.doppiozero.com/materiali/glittering/il-tempo-ritrovato-di-masha-e-orso. Accessed 25 Aug 2017.

Pavlov, I. P., & Anrep, G. V. (1927). *Conditioned reflexes; an investigation of the physiological activity of the cerebral cortex*. London: Oxford University Press.

Ricœur, P. (1990). *Soi-même comme un autre*. Paris: Seuil.

Surace, B. (2017). Baby Simulacra. Semiotica dei cuccioli al cinema come incubatori di assiologie in Mangiapane F. (ed) in E/C n. 21 (forthcoming).

The Human-Animal Relationship and the Musical Metaphor in *The Great Animal Orchestra* by Bernie Krause

Emiliano Battistini

Abstract Taking as case study the discourse of *The Great Animal Orchestra*, the following paper presents the sociosemiotic analysis carried out on the work by Bernie Krause, focusing especially on the analysis of the CD *The Great Animal Orchestra Symphony*, of the exposition *Le Grand Orchestre des Animaux* and of the related Internet sites, which followed his first wider public presentation *The Great Animal Orchestra: Finding the Origins of Music in the World's Wild Places*. Through the musical metaphor, Krause's discourse calls again into question the human-animal relationship, reversing the common meanings – this discovered animal culture stands opposite to contemporary human barbarity – and approaching multi-naturalistic positions, proposing a new politics of nature. This became possible because he changed his point of view on animals listening to them, generating a new approach to the relationship human-animals as well as showing new characteristics of the animals themselves.

1 *The Great Animal Orchestra* as a Discourse

The Great Animal Orchestra is the title of the work by Bernie Krause (2012a, b), musician and naturalist. For over 50 years, he recorded the soundscape (Schafer 1977) of wild regions all over the world.[1] Thanks to the analysis of his recordings, translated in visual scores through spectrogram technique, Krause finds out that different animal species of the same habitat vocalise following precise rules, that is – according to the so-called niches theory – each parting the frequency fields in which

[1] The term "soundscape" stresses the totality of sounds that are present within an environment, urban, rural or natural. The English neologism *soundscape* comes from the term *landscape*, and it has been used and made famous by Schafer (1977) and by his colleagues of the *Word Soundscape Project*, with the aim to highlight the importance of the sonic dimension of the environment. Starting from Schafer's work, to whom his book is dedicated, Bernie Krause (2012a) examines particularly the "wild" soundscape, such as the non-anthropic soundscape.

E. Battistini (✉)
University of Palermo, Palermo, Italy
e-mail: emiliano.battistini@unipa.it

G. Marrone, D. Mangano (eds.), *Semiotics of Animals in Culture*,
Biosemiotics 17, https://doi.org/10.1007/978-3-319-72992-3_15

to communicate, in the very same way musical instruments carve out their specific "acoustic territory" inside the orchestral arrangement. Therefore, the wild soundscape is not based on chance, but it follows its own internal organisation, elevating the whole range of animal vocalisations to the level of music – the art of organised sound – and opposing them to the ever more invasive and disorganised cacophony of sounds produced by modern humankind. Through this musical metaphor, Krause's discourse calls again into question the human-animal relationship, reversing the common meanings – this discovered animal culture stands opposite to contemporary human barbarity – and approaching multi-naturalistic positions (Descola 2005), proposing a new politics of nature (Latour 1999). This became possible because he changed his point of view on animals listening to them, generating a new approach to the relationship human-animals as well as showing new characteristics of the animals themselves (Despret 2002).

Taking as case study the discourse of *The Great Animal Orchestra*, the following paper presents the sociosemiotic analysis carried out on the work by Bernie Krause, focusing especially on the analysis of the CD (Blackford and Krause 2014a) and of the exposition (Fondation Cartier 2016) following his first wider public presentation *The Great Animal Orchestra: Finding the Origins of Music in the World's Wild Places* (Krause 2012a), whose analysis has been already presented in another paper (Battistini 2017).

The totality of these different texts (book, CD, exposition, Internet sites, online videos, documentaries, etc.) has different expressive substance but a highly coherent content and becomes a real discourse within the wider context of Western science, culture and public opinion. To analyse from a semiotic point of view the intertextuality concerned by *The Great Animal Orchestra*, I will consider the concept of discourse (Greimas and Courtés 1979, p. 86; Marrone 2001) read as a methodological and theoretical tool that can take into account how the same cultural theme disseminates through different expressive substances which create a system inside the society. For this reason I choose a sociosemiotic approach, that is able to account at the same time for the shifting of meaning between the different discursive genres involved by *The Great Animal Orchestra* (scientific, ecologic, musical, artistic, etc.) and for the local effects of meaning of its various texts, inside a more general perspective of semiotics of culture (Fig. 1).

2 Visual and Musical Translation of the Animal Orchestra Metaphor

The mixture between animal sounds and humans sounds hoped for by Bernie Krause (2012a) – that we can learn from our ancestors as from native populations and that marks in a complex way the union between nature and culture inside Western society – is experimented directly by Bernie Krause and Richard Blackford in the work *The Great Animal Orchestra, Symphony for Orchestra and Wild*

Fig. 1 The poster of The Great Animal Orchestra symphony

Soundscapes, available in CD (Blackford and Krause 2014a). This mixture is realised both through the musical discourse of the work and through the graphics chosen for the cover of the CD.

For example, in the graphics of the CD cover, on the figurative level, we can recognise a big musical staff with some black figures drawn over it. On the left, a little bit after the initial bar of the staff, a tree with a big root shaped as a comma grows to the top dividing itself in three branches and creating a light crown; from the centre of the tree, a series of birds fly to the foliage following an ordered line shaped as a spiral; the latter stands on one of the little branches at the base of the crown. The interplay of the plastic categories, topologics (the central position of the tree), chromatics (the black as the specific colour of the musical scores) and eidetics (the peculiar continuous line created in this way, which becomes wider from the central spiral, grows to the top and comes back down taking the shape of a straight line cutting the spiral and ending with a curl), leads us to recognise in this graphic composition the figure of a treble clef. The figural level (Polacci 2012), characteristic of the treble clef, is suitable for this iconic play between plastic level and figurative level. On the right, in a similar way and following the same graphic strategy, some chimpanzees in a row descend the staff hanging onto its lines as they were branches or lianas. They are discrete entities on the plastic level, disposed in sequence, in black colour, on different points of the staff, sometimes they are placed on the lines and sometimes on the white holes between the black lines. This gives the figures a second figurative reading, since we can recognise them as notes, specific ones: B, A, E and D. The treble key composed by the tree and the birds and the notes-chimpanzees, both placed on the staff, stand out on a general coloured background: an intense orange colour blends from the bottom to the top becoming always more faint and leaving always more space to the blue that occupies the upper part of the drawing. Also here the plastic level, especially thanks to the chromatic and topologic categories, suggests us a figurative reading: it is possible to recognise a sky at dawn or at sunset. This figure is strengthened by the previous knowledge that we have about the practices of wild soundscape recordings: we know that the animals and especially the birds vocalise much more during the rising and the setting of the sun. The specific visual configuration created by the interplay of these different levels of depth (black figures, staff and background) translates the animal orchestra metaphor in visual form as general effect of meaning. The animal vocalisations are conceived in a musical way: the animals are voices (notes and treble key) inside a big natural (the background as a sky) composition (staff). The poster of the symphony (Fig. 1) integrates this visual metaphor creating a score for an ensemble of three voices, showing under the staff with the treble key or G key, a second staff characterised by a C key shaped as a queen bee and worker bees-notes and a third staff with a bass key or F key shaped as a shell and tiger-notes. The three staffs, gathered together in a score for three voices ensemble, create a visual synecdoche since they replace the whole orchestral score composed by many more voices, specific of the symphony.

As already showed by the CD cover graphic, we find again this interplay between plastic and figurative categories in the recorded music, thanks to the figural level that works as a translation between the plastic and the figurative one. The five movements that compose the symphony are based on a continuous dialogue between the symphonic part composed by Blackford and the wild soundscapes recorded by Bernie Krause. This dialogue is made by cross references played on the phonic categories (pitch, intensity), rhythmic (metric, duration) and timbric ones (nature, grain) (Jacoviello 2012, p. 212). Often the two parts, electroacoustic and instrumental, fit together according the orchestral principle, distributing the different acoustic niches from time to time: for example, if the electroacoustic part covers the high frequencies with insect songs or birdsongs, the instrumental part works on the middle and bass frequencies with cellos and basses and vice versa. *Introduction and Tuning* recreates at the same time a natural environment and a musical one: after a gibbon serenade, the orchestra recreates the early morning wild soundscape thanks to a chattering and shimmering texture ranging from high piccolos to the rumble of basses and contrabassoon, reclaiming the most recognisable soundscape patterns and playing with them. A theme emerges over a repeated five-chord sequence, and the music "travels" through a thunderstorm. The movement ends with downward swooping textures combining Arctic seal vocalisation, followed by the song of a humpback whale that substitutes the note of the flute that started the movement. In

Scherzo and Riffs, a chorus of Pacific tree frogs introduces the orchestral rhythmic session. Gourd, shakers, rattles and woodblocks extend the pattern and a syncopated xylophone rattles over the percussions. A piliated woodpecker punctuates the acoustic space, answered by a cascade of temple blocks.[2] The rest of the orchestra joins in the jam session, growing to a huge tutti extended riff. In *Elegy*, the chilling and microtonal chorus of howling wolves is supported by four French horns in contrapuntal clusters. The atmosphere is bleak, and the nature is harsh: a Minnesota beaver utters forlorn cries at the destruction of its family and habitat. A solitary bassoon extends the beaver's lament into a universal elegy for animals displaced by the mankind. In *March and Charge*, a chattering woodwind phrase is answered by the lowest instruments in the orchestra and the sub-bass murmurs and growls of African elephants. Over a soft orchestral tutti accompaniment, the horns play a melody associated with the massive but gentle elephant herd, followed by the sounds of a gorilla munching bamboo and chest beating. An energetic, fast texture with overlapping ostinato develops into a tense tutti punctuated by wind and brass stabs and the scream of an adult gorilla. In *Song of the Musician Wren*, the musician wren repeats a perky, angular enchanting tune and the orchestral wind answer in variations, with different harmonisations and rhythmic extensions of the song. The common potoo, with its perfect pentatonic call, interrupts the wren. But the wren's song develops and keeps going even despite the attempts of the screaming piha to drown her out. The motto theme re-emerges as the finale concludes with orchestra and birds joining together in a joyous celebration of natural and musical sound.

The Internet site of the project (Blackford and Krause 2014b) recalls the graphics on the CD cover and on the poster and offers a series of contents that explains its realisation and diffusion. Analysing the sections of the site, in addition to the classic ones about contacts, the online shop and its presentation, we find a description of the musical section focused on the character of Richard Blackford, a scientific section related to the bioacoustics of Bernie Krause and, finally, a section dedicated to animals, which presents some of the animals who "sing" in the symphony thanks to the field recording. After a short explanation concerning the use of spectrograms – where it is claimed that the ones generated from animal vocalisations are very similar to musical score – for every animal (humpback whale, wolves, musician wren, Bornean gibbons), we find the name of the species, a picture, a recording of the song and its spectrogram. If on one hand the animals climb to the degree of protagonists of the symphony as the authors Blackford and Krause, on the other hand, the strategy of personalisation is not accomplished, presenting us an anonymous specimen for every species. However, the focus is to present examples of different animal species vocalisations: in order to listen to other examples, we are invited to visit the personal Internet site of Bernie Krause (1968), wildsanctuary.com.

[2] The *temple blocks* (or *blocks* of Korean woods) are musical instruments made by empty wood block, and they are used to play rhythmic impulses, especially in China, Japan and Korea during religious events. It is possible to play them by hands or leaned on some cloth.

3 *Le Grand Orchestre des Animaux*: An Immersive Installation

If the Internet site of *The Great Animal Orchestra Symphony* presents a traditional modality of scientific representation of animal vocalisation, divided by species and collected one by one according to the method of collection of fragments, the video and sound installation *Le Grand Orchestre des Animaux*, heart of the exposition of the same name presented at the Fondation Cartier pour l'art contemporain of Paris, between July 2, 2016, and January 8, 2017, realises an immersive listening situation (Fig. 2), much more close to the philosophy that supports Krause's work (Krause 2012a).

The rectangular room located in the museum basement floor, set up by the collective of London artists *United Visual Artists,* presents on the right a neutral wall, a passage and a series of seats, while on the left, a whole wall screen occupies the space up to the middle of the entrance wall and up to the middle of the opposite wall. This spatial configuration suggests to turn on the left to the wall screen theatre and to sit on the arranged seats. The room is dark, and the only light comes from the images that flow along the three screens, while the sound comes from hidden speakers placed under the walls and on the ceiling in a scattered and uniform way. The audio-video installation is composed by a long loop of about 1 h and a half that allows to listen to seven wild soundscapes recorded by Krause in different places of the world and, at the same time, to watch in synchrony at the relative spectrograms projected on the walls. Every recording is introduced by a short description of the place in which it has been recorded, with information about the geographic position, the weather, the date and the hour. When sound starts, a unique visual system launches on the screens: on the left, recorded sound frequencies and intensity appear and disappear in the form of straight horizontal lines, depending on the appearance

Fig. 2 Le Grand Orchestre des Animaux immersive installation

or disappearance of their relative sounds. This frequency movement is fixed on the central screen, the widest, gradually creating the recording spectrogram that starts to flow slowly from left to right. On this spectrogram, some information about the sonic sources that we are listening to are visualised from time to time, in a punctual mode: the attention of the viewer/listener is caught by little words that appear on the screen stressing the frequency fields with the most interesting form and consistency, specifying the species' name of the animal that is vocalising in that moment and the frequencies number of its song. Subsequently, these numeric and verbal information flow to the right, entering in the last screen without any modification and finally disappearing followed by other images. Every dynamic spectrogram related to its specfic wild soundscape plays on a different but limited colour palette on black background: the colours that recur are especially grades of red, green and blue, which alternate one another from recording to recording. As a general effect of meaning, this specific dynamic syncretism between visual score and musical score creates, on one hand, a progressive aestheticism of the spectrogram related to the scientific discourse and, on the other hand, a scientific frame to the sonic fascination and aesthetic of recordings. Scientific discourse and aesthetic discourse mix together on the basis of a general pedagogic discourse that aims to make the public aware about acoustic ecology. In fact, all this audio-visual narration is introduced by a video in which we see Bernie Krause, in a black and white medium close-up shoot, looking at the public through the central screen and speaking about his work and its usefulness. In this way, using the first person, the strategy on the enunciation level (Greimas and Courtés 1979, p. 104) of the video is coherent with that of the book (Krause 2012a). The length of the recordings and the time of the listening, the controlled and diffused intensity of sounds, the visual landscape created by the screens and the large space of the room set up with carpet and comfortable seats, all these elements create such an immersive fruition that the public, rather than using the seats as proposed by the communicative contract of the room through its spatial organisation, tends to lie down using the seats as seatbacks or pillows.

If *Le Grand Orchestre des Animaux* installation is coherent with the previous work by Krause, the rest of the exposition, composed around his work, is less consistent both in contents and forms. The theme of the great animal orchestra is developed through several animal figures, very different both about represented characters and aesthetics and expressive languages used: draw on paper (Cai Guo-Qiang), painting (Pierre Bodo, Moke, JP Mika, Cyprien Tokoudagba), photography (Manabu Miyazaki, Hiroshi Sugimoto, Christian Sardet), video (Les oiseaux artistes), painted ceramic (Adriana Varejao) and installation (Agnès Varda, Shiro Takatani, Ryuici Sakamoto). The semantic coherence is restored on the architectonical level: the different artistic works are placed following the scenography and the spatial organisation proposed by Mauricio Rocha and Gabriela Carrillo. In fact, the two architects have reorganised the space of the ground floor of the Fondation Cartier through the circular shape of the orchestra, creating semicircular sections of different dimensions in accordance with the exposed works which take the place of orchestra players. The circular scenography absorbs the glass walls of the foundation and goes beyond them including part of the garden and the entrance of the building (Fig. 3),

Fig. 3 The entrance of Le Grand Orchestre des Animaux exposition and the glass building by Jean Nouvel for the Fondation Cartier pour l'art contemporain

which welcomes the public as they were in the centre of the orchestra theatre, at the place of the conductor. The building is suited for this scenography: designed by Jean Nouvel for the Fondation Cartier pour l'art contemporain, it is known for its unique use of the glass walls that is possible to find also in other "new museums" designed by the same author such as the Quai Branly museum in Paris (Pezzini 2011, p. 111–114). The transparency creates variable limits between internal and external space, between exposition rooms and garden, between little and big scale, allowing an architectonical and artistic intervention that plays with the interstitial spaces specific of the orchestra. All is gathered and unified inside the same collective orchestra (exposition and artworks, garden, street, city, public, sky, etc.), overcoming the semantic opposition between nature and culture, as hoped for in *The Great Animal Orchestra* by Bernie Krause. Therefore we understand that besides the great animal orchestra, a "big orchestra of artists" is also presented. Every artist contributed to the general animal representation with his/her own voice, personality and expressive modality. Unfortunately, the harmony that we find in the wild soundscapes by Krause is not present on the visual level, and as a general effect of meaning, the great artist orchestra tends much more to the cacophonic than to the harmonic, lacking the translation on the visual level of that ordered division of acoustic frequencies characteristic of wild soundscapes.

4 Looking for the Conductor

A much more coherent effect of meaning is given by both the Internet site of the exposition (Krause, Fondation Cartier 2016) and its catalogue (Fondation Cartier 2016), both focused specifically on Bernie Krause's scientific and artistic work. The Internet site legrandorchestredesanimaux.com opens with a video presentation of Krause and his work in first person, in line with the enunciative strategy both of the

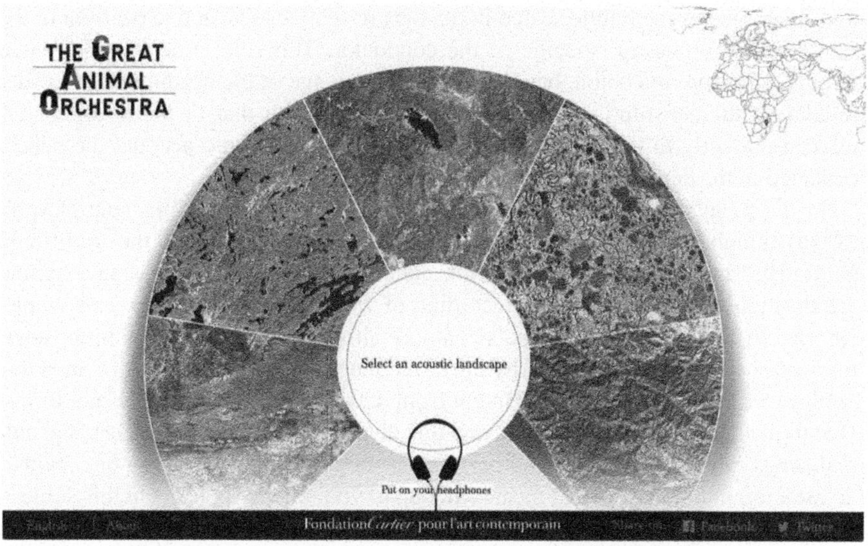

Fig. 4 A screenshot by Le Grand Orchestre des Animaux Internet site

book (Krause 2012a) and of the audio-video immersive installation. From the beginning, the site suggests to the user to wear headphones to listen to the soundscapes that in the meanwhile emerge in the foreground. The suggestion to wear the headphones is related to Krause's pedagogic discourse: as in the book the author gives the reader some basic concepts about audio technics and acoustics, in the same way we can re-experience partly Krause experience through the website, wearing the headphones (as he does when he is recording in the forest) and reading information about the habitats of which we are listening the soundscape. After the presentation, we can choose to listen to five examples of wild soundscapes clicking on the orchestral sections of a round figure that we recognise as a stylized orchestra where we are inserted – thanks to the headphone icon – at the place of the conductor (Fig. 4). At the same time, we can identify the same round figure as a CD which lacks the lower slice. The Internet site, on one hand, remediates (Bolter and Grusin 1999) in this way the CD player and, on the other hand, uses the same circular shape to visualise the orchestral theatre, inside whose sections the animals-musicians are inserted, through the recordings of their ensemble vocalisation: this is another way to visualise the great animal orchestra metaphor, this time using the figure of the CD rather than that of the staff. Each section is represented covered by a satellite picture of a different wild region. Clicking onto the images, we are directed to a specific page where we can listen to an editing or an excerpt of the relative soundscape, after a short description given us by the recorded voice of Krause. If, on the visual level, the graphic of the site uses diagrams and infographics, the hearing level made by the recordings listened through headphones creates a strong effect of reality and sonic iconism (Spaziante 2007, pp. 58–62). Through the Internet website, as in the spatial organisation of the exposition, the user passes from a distant position – the traditional one of the public seated in the theatre seats – to an active position in the centre of the orchestra, specific of the conductor. This role is never represented through a person or a baton, but always with the image of the headphones. In order to analyse the substitution of the conductor figure with that of the headphone – referring directly to the act of listening – we have to take into account the papers collected in the catalogue of the exposition.

In the essay *Figures de la re-composition*, the philosopher Vinciane Despret (2016) remembers the innovation proposed by Krause that left the traditional naturalistic methodology based on the collection of fragments of single animal vocalisation and approached the recording of wild animal soundscape as a whole. He tried to understand how animals compose altogether as well as within the environment, harmonising silences, rhythms, frequencies, etc., on the basis of an *active* regime of differentiation, very different from the one rather *passive* of visual forms. Due to this sharing of speech turns and this distribution in acoustic niches, the animal songs rarely overlap one another: it is as the animals were saying one another "c'est à toi, maintenant!" that means "it's up to you, now!" (p. 68). On this subject, Despret remembers Bruno Latour who, in the essay *La Société comme possession: La "preuve par l'orchestre"* (Latour 2011) whose title is taken from the film *Prova d'orchestra* (Fellini 1979)[3] by Federico Fellini, reconsiders the traditional idea – commonly accepted – of the politic organisation as a social contract based on the possession. According to this social contract theory, the society establishes itself cutting the space of goods in property exclusive zones. In the perspective of Hobbes, the division of property is necessary to avoid the "everybody against everybody" specific of the natural state. In this sense "c'est à toi, c'est à moi" means "it is yours, it is mine", on the basis of a spatial metaphor: it founds the political gesture, as a way of breaking up with nature and detaching from it. But Latour reflects that the "c'est à toi, c'est à moi" can be understood also with the meaning of "it is up to you, it is up to me", on the basis of a temporal metaphor, specific of the orchestra situation in which it is necessary to mark a common action, coordinating our own part with the others, inside a general relationship of interdependence. The regime of politic composition based on this updated definition of possession would not be a break with a natural state anymore but rather a re-creation of it. On one hand, this marks the limit of the human political compositions – when humans compose, i.e. organise themselves collectively excluding non-human beings. On the other hand, this new definition of possession leads to reconsider the way through which humans speak about and represent animal society. For example, speaking about "troops of baboons", "hordes of insects" and "primitive violence", the politic models proposed by several ethologists and primatologists show a recurrent organisation scheme as a hierarchical type in which the domination is exercised from the top to the bottom and where only the favoured can make their voice to be heard. This repossessing of one of the possible forms of politic composition evoking the natural orchestra opens

[3] I thank Isabella Pezzini for reminding me that also in the film *La dolce vita* (Fellini 1960) by Federico Fellini, there is a character that records sounds from the environment, but in this case, it is connected to the category of disturbing.

the imaginary to unexpected forms and modalities that broke with the models of possession as property and of the hierarchic power. We can now take into consideration the role of the conductor. Always Despret, following Latour thought, stresses that all the work of the conductor is to break the routine of the orchestra players, to ensure that they are at the same time in their part and out of it, that they are interested to others staying in their role, because "if everybody were aware only of his/her part and only of that or if would limit himself/herself to his/her role and only of it there would not be society neither orchestra" (Latour 2011, p. 13).[4] Then the conductor does not impose but remember to listen to others. The metaphor of the orchestra insists on the necessity of a well-regulated act together, but especially it makes inevitable the necessity to pay attention to everybody, to our activity as well as to the others'. This is exactly what these interspecific sonic communities accomplish. They create networks of relationships and information that guide the way to operate, not necessarily together, but taking the others into consideration. The interest in the organisation and in the structure of wild soundscapes and the possibility to renegotiate the opposition between human and animal and between culture and nature by listening to them are shared also by the French composer François-Bernard Mâche:

> [...] The soundscapes are a way of hearing especially enriching since their contents, by their organisations, by their actors, everything seems to promote the search of a new relationship with nature. Today several sociologists, philosophers, prehistorians, ecologists advocate that the man is obliged to come down from his pedestal, too much shaken in different directions. It is a question of observing that culture is not radically opposed to nature exclusively in favour of our species, it rather forms itself inside the world of living things [...]. (Mâche 2016, p. 149)[5]

Coming back to the metaphor of the orchestra, as a composer, Mâche ends his paper with a provocation: "de facto, a great orchestra needs a leader. But so where is he? Hush… Answer later?" (Mâche 2016, p. 149).[6] We are asked on the curious lack of the figure of the conductor in the Krause discourse, compared to the rich discursive configuration (Greimas and Courtés 1979, p. 51) developed by his orchestra metaphor. Who or what occupy the place of the conductor in the great animal orchestra? the nature? the human being? The discourse of Bernie Krause remains ambiguous on this point. More than an ontological entity, it is possible to consider the place of the conductor as a structural entity: his place is that of an "empty case" (Deleuze 1972), to fill from time to time, contextually, as it happens to the public of the exposition or to the users of the Internet site of the *Le Grand*

[4] The translation is mine.

[5] [...] les paysages sonores sont une façon d'entendre d'autant plus enrichissante que par leurs contenus, par leurs organisations, par leurs acteurs, tout semble favoriser la recherche d'un nouveau rapport avec la nature. Plusieurs sociologues, philosophes, préhistoriens, écologues plaident aujourd'hui pour constater que l'homme est. bien obligé de descendre de son piédestal, décidément trop secoué dans divers sens. Il s'agit de constater que la culture ne s'oppose pas radicalement à la nature au seul profit de notre espèce, et qu'elle s'esquisse dans le monde du vivant. [...] (Mâche 2016, p. 149). The translation is mine.

[6] The translation is mine.

Orchestre des Animaux, in which the spatial organisation, architectonical and digital, leads the listener to the centre. The empty case is that one related to listening: everyone is able to be in the best listening place, that of the conductor, who can listen to all the voices in a balanced way and at the same time. A vacant place that is filled at one's own turn, it enables the continuous play of the meaning and the continuous and discursive redefinition of the human-animal and culture-nature relationships as well as of their overcoming towards a higher human collective order. As the listening to Bernie Krause's wild soundscape aims to suggest.

References

Battistini, E. (2017). La Grande Orchestra degli Animali. Un'analisi sociosemiotica. In G. Marrone & A. Giannitrapani (Eds.), *Zoosemiotics 2.0. Forme e politiche dell'animalità, Palermo, december 2016. I Nuovi Quaderni del Circolo Semiologico Siciliano* (Vol. 1, p. 489). Palermo: Museo Internazionale delle Marionette – Antonio Pasqualino.

Blackford, R., & Krause, B. (2014a). *The great animal orchestra. Symphony for orchestra and wild soundscape (CD)*. Nimbus Record, UK

Blackford, R., & Krause, B. (2014b). *The great animal orchestra symphony*. http://www.thegreat-animalorchestrasymphony.com. Accessed 4 Sept 2017.

Bolter, J. D., & Grusin, R. (1999). *Remediation*. Cambridge, MA: The MIT Press. Italian edition: Bolter, J. D., & Grusin, R. (2002). *Re-mediation*. Milano: Guerini.

Deleuze, G. (1972). A quoi reconnaît-on le structuralisme? In F. Chatelet (Ed.), *Histoire de la philosophie. Le XX siècle* (Vol. 8, pp. 299–335). Paris: Hachette. Italian edition: Deleuze, G. (2004). *Lo strutturalismo*. Milano: SE.

Descola, P. (2005). *Par-delà nature et culture*. Paris: Gallimard.

Despret, V. (2002). *Quand le loup habitera avec l'agneau*. Paris: Les Empêcheurs de penser en ronde/Seuil.

Despret, V. (2016). Figures de la re-composition. In Fondation Cartier pour l'art contemporain (Ed.), *Le Grand Orchestre des Animaux, The great animal orchestra* (pp. 67–74). Paris: Fondation Cartier pour l'art contemporain.

Fellini, F. (1960). *La dolce vita (film)*. ITA-FR, 180′.

Fellini, F. (1979). *Prova d'orchestra (film)*. ITA, 90′.

Fondation Cartier pour l'art contemporain. (2016). *Le Grand Orchestre des Animaux, The great animal orchestra, catalogue of the exposition*. Paris: Fondation Cartier.

Greimas, A.-J., & Courtés, J. (1979). *Sémiotique. Dictionnaire raisonné de la théorie du langage*. Paris: Hachette. Italian edition: Greimas, A.-J., & Courtés, J. (2007). *Semiotica. Dizionario ragionato della teoria del linguaggio*. Milano: Mondadori.

Jacoviello, S. (2012). *La rivincita di Orfeo. Esperienza estetica e semiotica del discorso musicale*. Milano-Udine: Mimesis.

Krause, B. (1968). *Wild sanctuary*. http://www.wildsanctuary.com. Accessed 4 Sept 2017.

Krause, B. (2012a). *The great animal orchestra. Finding the origins of music in the world's wild places*. New York: Little Brown and Company.

Krause, B. (2012b). *The great animal orchestra*. http://www.thegreatanimalorchestra.com. Accessed 4 Sept 2017.

Krause, B., & Fondation Cartier pour l'art contemporain. (2016). *Le Grand Orchestre des Animaux*. www.legrandorchestredesanimaux.com. Accessed 4 Sept 2017.

Latour, B. (1999). *Politiques de la nature*. Paris: La Decouverte & Syros.

Latour, B. (2011). La Société comme possession. La 'preuve par l'orchestre. In D. Debaise (Ed.), *Philisophie des possessions*. Dijon: Les Presses du Réel.

Mâche, F.-B. (2016). Mais où est donc le chef d'orchestre? In Fondation Cartier pour l'art contemporain (Ed.), *Le Grand Orchestre des Animaux, The great animal orchestra, catalogue of the exposition* (pp. 143–150). Paris: Fondation Cartier pour l'art contemporain.

Marrone, G. (2001). *Corpi Sociali. Processi comunicativi e semiotica del testo*. Torino: Einaudi.

Pezzini, I. (2011). *Semiotica dei nuovi musei*. Roma-Bari: Laterza.

Polacci, F. (Ed.). (2012). *Ai margini del figurativo*. Siena: Protagon.

Schafer, R. M. (1977). *The soundscape: Our sonic environment and the tuning of the world*. New York: Knopf.

Spaziante, L. (2007). *Sociosemiotica del pop. Identità, testi e pratiche musicali*. Roma: Carocci.

The manufacturer's authorised representative in the EU is Springer Nature Customer Service Centre GmbH, Europaplatz 3, 69115 Heidelberg, Germany. If you have any concerns regarding our products, please contact ProductSafety@springernature.com

Printed and bound by CPI Group (UK) Ltd, Croydon, CR0 4YY
30/09/2026
02225587-0001